Digital Genres in Academic Knowledge Production and Communication

STUDIES IN KNOWLEDGE PRODUCTION AND PARTICIPATION

Series Editors: **Mary Jane Curry**, *University of Rochester, USA* and **Theresa Lillis**, *The Open University, UK*

Questions about the relationships among language and other semiotic resources (such as image, film/video, sound) and knowledge production, participation and distribution are increasingly coming to the fore in the context of debates about globalisation, multilingualism, and new technologies. Much of the existing work published on knowledge production has focused on formal academic/scientific knowledge; this knowledge is beginning to be produced and communicated via a much wider range of genres, modes, and media including, for example, blogs, wikis, and Twitter feeds, which have created new ways of producing and communicating knowledge, as well as opening up new ways of participating. Fast-moving shifts in these domains prompt the need for this series which aims to explore facets of knowledge production including: what is counted as knowledge, how it is recognised and rewarded, and who has access to producing, distributing and using knowledge(s). One of the key aims of the series is to include work by scholars located outside the 'centre', and to include work written in innovative styles and formats.

All books in this series are externally peer-reviewed.

Full details of all the books in this series and of all our other publications can be found on http://www.multilingual-matters.com, or by writing to Multilingual Matters, St Nicholas House, 31-34 High Street, Bristol, BS1 2AW, UK.

STUDIES IN KNOWLEDGE PRODUCTION AND PARTICIPATION: 4

Digital Genres in Academic Knowledge Production and Communication

Perspectives and Practices

María José Luzón and Carmen Pérez-Llantada

MULTILINGUAL MATTERS
Bristol • Jackson

DOI https://doi.org/10.21832/LUZON4719
Library of Congress Cataloging in Publication Data
A catalog record for this book is available from the Library of Congress.
Names: Luzón, María José, author. | Pérez-Llantada, Carmen, author.
Title: Digital Genres in Academic Knowledge Production and Communication: Perspectives and practices / María José Luzón and Carmen Pérez-Llantada.
Description: Bristol, UK; Jackson, Tennessee: Multilingual Matters, 2022. | Series: Studies in Knowledge Production and Participation: 4 | Includes bibliographical references and index. | Summary: "This book presents an overview of the wide variety of digital genres used by researchers to produce and communicate knowledge, perform new identities and evaluate research outputs. The book explores what researchers can do with these genres, what meanings they can make and what language(s) they deploy in carrying out all these practices"—Provided by publisher.
Identifiers: LCCN 2021053174 | ISBN 9781788924719 (hardback) | ISBN 9781788924733 (epub) | ISBN 9781788924726 (pdf)
Subjects: LCSH: Education, Higher—Effect of technological innovations on | Language and languages—Computer-assisted instruction. | Knowledge management. | Communication in learning and scholarship.
Classification: LCC LB2395.7 .L89 2022 | DDC 378.1/734—dc23/eng/20220124
LC record available at https://lccn.loc.gov/2021053174

British Library Cataloguing in Publication Data
A catalogue entry for this book is available from the British Library.

ISBN-13: 978-1-7889-2471-9 (hbk)

Multilingual Matters
UK: St Nicholas House, 31-34 High Street, Bristol, BS1 2AW, UK.
USA: Ingram, Jackson, Tennessee, USA.

The policy of Multilingual Matters/Channel View Publications is to use papers that are natural, renewable and recyclable products, made from wood grown in sustainable forests. In the manufacturing process of our books, and to further support our policy, preference is given to printers that have FSC and PEFC Chain of Custody certification. The FSC and/or PEFC logos will appear on those books where full certification has been granted to the printer concerned.

Typeset by Deanta Global Publishing Services, Chennai, India

Contents

Figures

Tables

Acknowledgments

This book is a contribution to the project 'Digital Genres and Open Science', funded by the Spanish Ministry of Science and Innovation (PID2019-105655RB-I00/ AEI/10.13039/501100011033). It is also a contribution to the research carried out within the research group Comunicación Internacional y Retos Sociales (CIRES) funded by the Government of Aragon (project code H16_20).

Permission Acknowledgements

We would like to acknowledge the following copyright holders for permission to reproduce the following material:

- John Benjamins for the reproduction of the figure 'A genre-theoretical model that includes vertical (supergenres vs. subgenres)' in Figure 2.1. Adapted from Heyd (2008). *Email Hoaxes: Form, Function, Genre Ecology* (https://benjamins.com/catalog/pbns.174)
- *Faculty Opinions* (https://facultyopinions.com/) for permission to reproduce a screenshot of recommendations for the paper 'Words without meaning' (Figure 10.1)
- *Experiment.com* for permission to reproduce screenshots from their website (Figures 8.1 through 8.4)

We would also like to thank the following researchers who have granted us permission to use images from their digital documents:

- Members of the GENTIC group for permission to use a screenshot of the post "PREZIosa initiative" (Figure 4.1)
- Rodolfo Garcia-Contreras and Jesús Guillermo Jiménez-Cortés for permission to reproduce an image from their crowdfunding project in Experiment.com (Figure 8.1)
- Mario Williams and Jacquelyn Gill for permission to reproduce images from their crowdfunding project in Experiment.com (Figures 8.2 through 8.4)

- Jesús Campos Group for permission to reproduce two tweets from their Twitter account (Figures 7.1 and 7.2)
- GT&MA Research Group for permission to reproduce a tweet from their Twitter account (Figure 7.3)
- SOLTI Group for permission to reproduce two tweets from their Twitter account (Figures 7.4 and 7.5)
- Blanca Ros, Jorge Cored, Francisco Serón and Bernardino Moreno for permission to reproduce screenshots from the videos *Liquid Crystals* (Figure 9.2), *Virtual Archaeology* (Figure 9.3) and *The Journey of the Steak* (Figure 9.4)

1 Introduction: Why Focus on Digital Genres?

1.1 Introduction

The unique technological affordances of new media and the impact of open science are revolutionizing the way researchers worldwide produce, represent, reuse and share knowledge. Although the status of the research article as the primary genre in academic knowledge[1] production and communication remains uncontested, the affordances of the internet (e.g. hyperlinking, multimodality and interactivity) provide researchers with a rich repertoire of digital genres to share, at different moments in time, a variety of data types and results with diverse audiences, widen their participation in the scholarly arena and interact with the lay public.

In a context of considerable socioeconomic, cultural and technological change, researchers need to meet new and more complex demands. In addition to publishing high quality research, scholars in many contexts are required or urged to promote their research output and achieve visibility and international impact, in order to meet institutional standards, and achieve personal recognition and public funding. We are often encouraged to participate in national and international networks and to collaborate with other researchers and research groups. Funding agencies also often demand more transparent and 'open science', that is, science that involves open sharing and dissemination of scholarly research to expert and diversified audiences. Thus, researchers are encouraged to establish a dialogue with the general public, engage citizens in research processes and disseminate knowledge to non-specialist readerships. We may also need to show that research methods are reliable and the data shareable, and to interact with peers and non-specialist audiences in different media. Digital genres for knowledge communication have emerged and evolved as scholars respond to these demands and adapt to the changing research and academic context. This book seeks to give a comprehensive view of how scholars employ digital genres to disseminate academic knowledge to diverse audiences in response to these social and economic demands. The book discusses how digital genres are changing the landscape of academic communication and brings to the fore important changes in scholars' communication practices; it thus illuminates the genre knowledges that scholars may need in order to achieve social and individual goals in this changing context as well as the challenges that they may face.

Genres, defined as semiotic tools for accomplishing actions or goals (Devitt, 2004; Miller, 1984; Miller *et al.*, 2018), have become crucial to our understanding of knowledge communication and dissemination. Looking at genre-mediated actions enables us to appreciate how the rapid technological advances of the internet shape and constrain the ways in which academic knowledge is produced and exchanged. In this book, we use the term 'digital genre' to refer to internet-mediated genres, that is, genres which harness the affordances of the internet to varying degrees. It includes what we might think of as more static genres, or genres which do not allow readers to add or contribute content (e.g. academic homepages), as well as more dynamic, participatory, collaborative or interactive genres (e.g. online laboratory notebooks or enhanced publications).

Three concepts are central to an understanding of how digital genres facilitate and/or constrain researchers' communication practices today: medium, mode and language choice. Medium refers to the material substance through which the message is communicated. In the case of digital genres, this term signals any medium that encodes data and information which is stored and transmitted over the internet or computer networks. The specific technical features and capabilities of digital media enable actions often not possible using traditional media, for example, to reach both expert peers and broader publics simultaneously or to share research data and results free of charge, and before formal publication, with other researchers. For scholars, these capabilities not only expand the possibilities for communication and collaboration with their disciplinary community and for public outreach but they also impose some constraints and pose challenges, as we discuss throughout this book.

Particular media make specific modes available. A mode is a 'socially organized set of semiotic resources for making meaning', for example, writing, sound as speech, music, still and moving images (Jewitt *et al.*, 2016: 157). What is particularly relevant for the discussion of digital genres in this book is the fact that the different modes made available in digital genres have different potentialities and limitations for making meaning, that is, they have different 'mode affordances' (Kress, 2010). For instance, the affordances of moving images include the passage of time and movement (Kress, 2010). Both the medium and the semiotic resources that can be deployed in a genre shape the potentials and constraints of such genre for meaning making. As an example, the modes afforded in the video methods article (VMA) (e.g. moving image, spoken narration) (see Hafner, 2018) make it possible to demonstrate complex methods and laboratory techniques. Digital genres are inherently multimodal (Herring, 2019; Jones & Hafner, 2012), with written verbal text being only one of the possible semiotic resources through which meaning is made.

Finally, the language(s) that researchers choose when producing digital genres also offer(s) particular potentialities, which we consider as 'language affordances'. The digital medium enables researchers to engage in complex multilingual practices often not adopted in conventional academic communication (Androutsopoulos, 2015) and draw upon two or more languages that are part of their linguistic repertoire (i.e. English and/or the languages spoken in their local communities and/or languages for cross-border communication) in order to reach and connect with various audiences (i.e. international or local audiences).

1.2 Why this Book?

Although in the last few years there has been an increasing interest in the genres used by researchers to communicate knowledge online, existing studies have focused mainly on how particular genres are changing (in form and substance) because of the increasing use of digital platforms and multimedia elements, and on the purposes for which some digital genres are used by specific discourse communities (see e.g. Luzón, 2017; Rowley-Jolivet & Carter-Thomas, 2019; Wickman, 2016). However, in order to understand contemporary knowledge production and dissemination, we need a more comprehensive picture of the diversity of digital genres on which researchers draw to communicate within and beyond the boundaries of traditional discourse communities. It is important to understand the rhetorical actions performed by these genres, the multiple semiotic modes that researchers can combine when composing these genres, and the languages in which researchers based in non-Anglophone institutions produce these genres. This book thus aims to provide a holistic view of the wide variety of digital genres used by researchers, in an attempt to shed light on broader changes in academic communication in the 21st century (i.e. new purposes, new audiences and new academic knowledge production and dissemination practices). Our account should be of interest in itself but it is also of pedagogical value, since it helps researchers become aware of academic digital practices and of the skills they may need in order to produce and communicate knowledge in the digital era.

Specifically, we seek to answer four key questions in this book:

(1) What are the digital genres being produced by academic researchers and why have these genres emerged?
(2) What are the purposes of the digital genres used in current knowledge communication practices and what are their affordances and constraints?

(3) Who are the audiences of these digital genres and what rhetorical strategies are deployed to reach such audiences?
(4) What language(s) and semiotic resources are used in digital genres today and what is the rationale behind researchers' language choice?

The question of language is of particular interest for us, since we ourselves are multilingual scholars who do research in a European national context where English is not an official language of communication but is often an explicit and implicit requirement for our research practice. The choice of language for communicating knowledge is an important decision for scholars carrying out research in non-Anglophone research institutions around the world. We may use English in online communication as a lingua franca for reaching international and linguistically diverse audiences (Luzón, 2018b). English supports collaboration among researchers from diverse linguacultural backgrounds and enables them to share their results with the international expert community. On the other hand, as we will illustrate in this book, researchers may use their local/national languages or engage in multilingual practices for a number of reasons, including performing multiple identities or reaching diversified audiences. The languages used to communicate knowledge online have received little attention, with limited research on how multilingual scholars draw upon the linguistic codes that are available to us when communicating in online contexts or on the interaction between languages and digital genres. In this book, we hope to contribute insights into the languages used for communicating knowledge. We argue that in the digital media era, while the choice of a particular language or languages depends on the contextual specificity of genres and on the target audience the scholars aim to reach, multilingualism widens the possibilities for accessing, sharing and disseminating knowledge to diverse audiences. One goal of this book is to contribute to raising awareness of the importance of multilingualism in scholarly communication online, specifically in providing access to knowledge to diversified audiences.

The issue of audiences is therefore also particularly important when considering digital genres. In this book, we engage with the concept of 'context collapse' (boyd, 2002) and show that, although this concept has been critiqued (Szabla & Blommaert, 2020), it remains useful as a way of capturing how, due to the open nature of many of these genres, multiple audiences are often collapsed into a single context. The hybrid and flexible nature of many digital genres makes them powerful tools not only to communicate with peers but also for dissemination to the lay public, which enables a much broader participation in the scholarly process. This book will contribute to mapping the audiences with whom researchers communicate and collaborate in the 21st century and to showing how researchers adapt their discourse to reach and meet the needs of these audiences.

The analysis of digital genres for knowledge communication requires an awareness that genre emergence and change are not driven by technology alone, but are mediated by social, cultural, linguistic and material dimensions. As pointed out previously, these digital genres have emerged as responses to the changing research context and needs that scholars face. In addition, although they facilitate sharing, greater access and circulation, and more transparency and visibility, these genres also bring with them new challenges and raise broader questions related to their role in global academia (e.g. socioeconomic interests, issues of power, visibility and access). We aim to both describe the ways in which digital genres are being used and also critically assess the inherent limitations of these genres, the constraints imposed by digital media and by broader socioeconomic factors, and the challenges that they pose to researchers.

1.3 Why Digital Genres?

As applied linguists interested in the study and pedagogy of academic communication, we advocate for genre analysis as an appropriate and valuable theoretical and analytical framework for understanding knowledge communication in the digital context. We draw on both traditions of genre analysis from English for Specific Purposes (Bhatia, 2002, 2004; Swales, 1990) and Rhetorical Genre Studies (Devitt, 2004; Miller, 1984, 2015a), since both traditions are based on a functional view of genres that helps us to explain why digital genres emerge and what researchers do with them. Thus, following pioneering work on genres in these traditions, we conceptualize 'genres' as tools for responding to particular rhetorical exigences and achieving particular communicative purposes (Bawarshi & Reiff, 2010; Miller, 1984; Swales, 1990). Like Heyd (2016: 88), we argue that genre analysis is a suitable 'toolbox for doing linguistic and discourse analysis with digital material' because it provides a useful way not only to reveal the communicative purposes and effects of emerging forms of online communication and describe their discursive features, but also to explain why these forms emerge and evolve. As Bhatia (2002: 6) explains, genre analysis helps us 'to understand how members of specific discourse communities construct, interpret and use [these] genres to achieve their community goals and why they write them the way they do'. Since digital genres are fundamentally multimodal (Herring, 2019), the analysis of these genres involves studying how various multisemiotic resources are employed to achieve the purposes of the genre. A more detailed discussion of genre and the framework for digital genre analysis applied in this book is presented in Chapter 2.

The chapters in this book explore digital genres, focusing on what scholars do with them (see Figure 1.1).

Digital genres are used by scholars to perform, among other things, the following actions: to build their scholarly reputation and negotiate

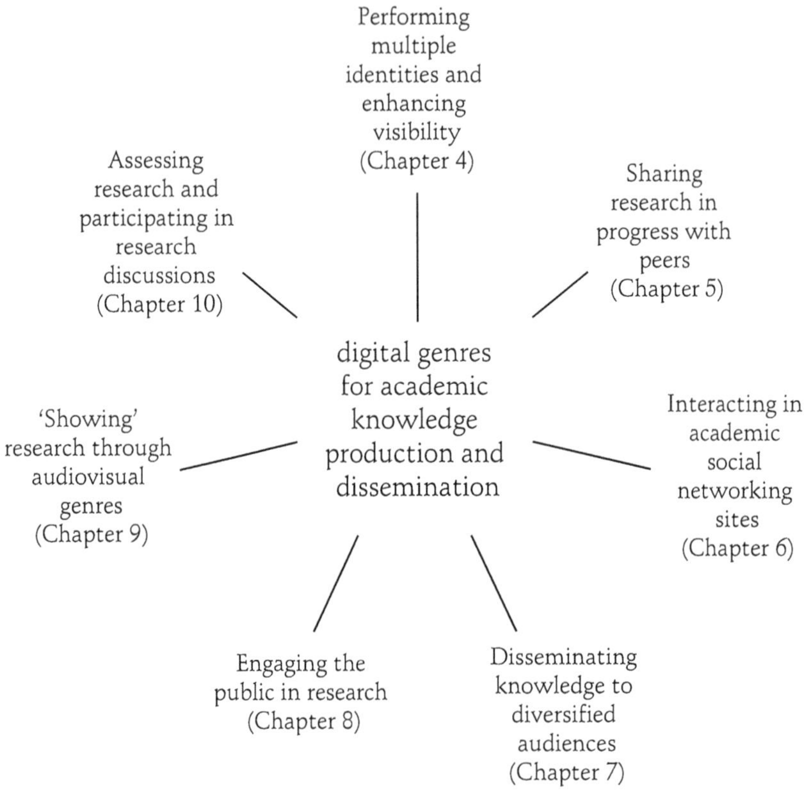

Figure 1.1 What do researchers do with digital genres?

their academic identities; to share research in progress and practices with peers; to interact with scholars across the world and form new relations with other researchers; to disseminate research to diversified audiences in informal contexts; to engage the public in funding or conducting research; to create meaning in new ways by combining various modes; and to assess the quality of research and provide feedback in new ways.

Chapters 4–10 discuss the use of digital genres to accomplish these actions. Each of these chapters first presents a critical synthesis of current research on different practices of knowledge production and dissemination in the digital era, followed by what we refer to as a 'research case study'.[2] Each case study involves investigating a small number of related exemplars of a specific digital genre within the shared context in which they are used. The primary data sources for these studies are therefore small-scale specialized corpora of digital genres. The genres that the studies focus on comprise research group blogs, open lab notebooks (OLNs), tweets by research groups, crowdfunding project proposals, posts on academic social networking sites (ASNSs), online science videos and science blogs. Given our interest in multilingualism, in some of the chapters we analyze genre exemplars produced exclusively by multilingual scholars (affiliated with Spanish universities), which include languages other than English (mainly Spanish, but also Catalan). The case studies illustrate what researchers do with these genres, what meanings they make, who they interact with, how they engage with their audiences interpersonally, what identities they construct and, finally, what language(s) they deploy in carrying out all these practices. Given our overarching genre framework, which involves viewing genres as complex and multilayered phenomena (Bhatia, 2002), analysis of each case involves paying attention to specific features and dimensions, always taking into account the multimodal nature of digital genres: we conduct rhetorical move analysis (in Chapters 5 and 7-9), analysis of discourse strategies and features (in Chapters 5-10) and content analysis (in Chapters 4 and 7). We hope that the exploratory case study approach we have adopted will reveal the particularities of the context and enable a better understanding of the existing and emerging communication practices in today's scholarly community.

1.4 Outline of the Book and Audience

The rationale for this book is to present an overview of existing digital genres that contribute to structuring and enabling research activity and thus show how the emergence of these genres and the new discursive practices they involve are changing the landscape of scholarly communication. In order to frame the analysis of digital genres within a theoretical perspective, in Chapter 2 we discuss the concept of genre, particularly in relation to knowledge communication on the internet. The chapter summarizes some issues and questions that have been raised in relation to digital genres, for example, how genres have evolved when migrating to the internet and emerged as a response to new social practices and demands; how Web 2.0 affordances are problematizing the concept of 'genre'; and the difficulty in distinguishing between genres and medium in this context.

Chapter 3 provides an introduction to the current affordances, challenges and constraints on researchers' activity in the 21st century and how they influence knowledge communication. The chapter explores: the increased globalization of research and the importance of international collaboration; the new social demands that researchers need to respond to; the blurring between scholarly discourse and public communication of science in recent years as a result of an increasing democratization of knowledge and of the demand for a more participatory framework; researchers' choice between (or combination of) English as an international language and other languages to reach diversified audiences; and the possibilities offered by Web 2.0 technologies for scholarly communication.

In Chapters 4–10, we present a critical synthesis of current research on different dynamics of academic knowledge communication and dissemination in the digital era. These dynamics are then illustrated and discussed in more detail through our own studies of specific digital genres. Chapter 4 discusses how digital genres with a primarily promotional purpose are used by scholars to construct multifaceted identities and increase their visibility. As insightfully stated by Bazerman (1997: 19), identity and genre are concepts intimately related: 'Genres are forms of life, ways of being'. Academic communication is an act of identity and the construction of a credible academic identity is of central importance to scholars, hence the need to analyze how digital genres are used for this purpose. The chapter focuses on identity construction in digital academic discourses as a situational and multimodal activity. The study in this chapter explores how Spanish research groups who write their blogs in English (or in English and in their local languages) use these blogs to negotiate multiple identities.

Chapter 5 is concerned with a digital genre used by researchers to share research in progress: OLNs. It first reviews research on OLNs, a genre that responds to the need to collaborate and share research results and protocols before their publication. The second part of the chapter presents the results of a rhetorical and discourse analysis of posts in a set of OLNs, which serves to describe the discourse practices that researchers engage when composing this genre and illustrates how the texts are strategically crafted to invoke solidarity and promote collaboration.

In Chapter 6, we discuss ASNSs as 'translocal affinity spaces' (Leppänen & Peuronen, 2012), that is, places where users from different linguacultural backgrounds and disciplines can share interests and concerns with other like-minded people. ASNSs such as ResearchGate or Academia.edu have gained popularity among scholars as spaces in which to publicize their research outputs and develop an academic identity. They are also used by academics to connect with other researchers, discuss research, exchange information and suggestions and receive feedback from experts in the same field. The case study in this chapter consists of an analysis of Q&A forums in ResearchGate, which illustrates how these ASNSs are used as forums for cross-disciplinary and transcultural communication, where scholars with different degrees of expertise not only share knowledge informally, but also get emotional support and help.

Chapter 7 focuses on how research centers and institutions, research groups and individual researchers use digital genres (e.g. Twitter, blogs) to communicate their research results and those of others to a diversified audience. The chapter explores genres used by researchers to interact in real time both with peers and the interested public, focusing on the different purposes and audiences of these genres and the interplay between audiences and languages. It discusses how these forums for interaction with the public can create public science engagement and facilitate the link between specific academic knowledge and the concerns of the lay public. In the last section, we present the results of an analysis of the Twitter accounts of several research groups affiliated with Spanish institutions, which shows how different semiotic resources are combined to achieve the groups' purposes and how language choice is used strategically to reach various audiences.

Chapter 8 is concerned with digital genres employed by scholars to facilitate the participation of the public in research. It discusses how researchers present their research in progress with interested publics in order to persuade them to collaborate by participating in the data collection process or by funding it. Taking the case of crowdfunding projects, the chapter illustrates how digital media support several semiotic modes and offer researchers various participatory tools and resources to educate the public on issues of science and persuade it to engage in research (public understanding of science and public engagement in science), by this means supporting the democratization of science, that is, making science accessible to lay publics.

In Chapter 9, we examine several (audio)visual genres that are employed to 'show' research and thus increase knowledge dissemination for diverse audiences: (audio)visual abstracts, which have emerged online to enhance the research article genre; the VMA, which has emerged to facilitate the replication of scientific protocols; and online science videos, used to engage both the expert and lay publics and to facilitate public understanding of science. In the last section, we examine how various semiotic resources are orchestrated in online videos produced by research groups to show their research to diverse audiences and to arouse the public's interest in and awareness of the importance of research in their discipline.

Chapter 10 analyzes digital genres whose main purpose is to assess research, focusing mainly on two aspects. The first is the emergence and evolution of genres for online peer review as a response to open science practices of quality control, and their potential impact on quality assessment processes. We discuss the online peer review of manuscripts submitted for publication, and the use of online tools to provide feedback on published research (e.g. comments and notes). The second related aspect is the use of blogs and microblogs to raise concerns about the validity or honesty of published research and to engage in scientific controversies. In the last section, we present the results of a research case study that analyzes how scholars use blogs to engage in the discussion of controversial disciplinary issues and shows the relation of these blog discussions to other scholarly genres used for academic criticism.

Lastly, Chapter 11 provides some final considerations about new genres for sharing disciplinary knowledge locally and globally. We critically assess the value of multilingual web-mediated academic communication practices with a view to postulating digital practices as an opportunity to support creative knowledge dissemination practices across disciplines. In closing, we suggest several directions for future research on digital genres for knowledge production and dissemination.

The book aims to appeal to researchers in the fields of academic communication, genre analysis, online communication and English for Academic Purposes (EAP), and student researchers who are attracted to research on communicating knowledge online in English as a lingua franca and in other languages. As an up-to-date review of practices and evolving strategies in knowledge communication, the book may also be helpful to academic researchers as they perform new roles and engage in new social practices. We hope that the book will provide a comprehensive account of the nature of digital genres for academic knowledge communication and of the ways these are opening up opportunities (and posing new challenges) for academic knowledge exchange with the multiple audiences with whom scholars need to communicate in the 21st century.

Notes

(1) In this book, we use the term 'academic knowledge' rather than 'scientific knowledge' (which is sometimes associated only with the natural sciences) to refer to scholarly knowledge produced across disciplines. However, we use the term 'science' to refer to some concepts that are not restricted to the natural sciences, such as 'open science', 'public understanding of science' or 'public engagement with science'. We also use the terms 'science' or 'scientists' when we report on research where these terms are used.

(2) We combine the notion of 'research case' (Flowerdew, 2012; Hyland, 2015b), to mean (small-scale) research projects which illustrate how texts and their contexts can be analyzed, with the notion of 'case study', which emphasizes that these studies have a narrow and specific focus of interest and exploration.

2 Genre as a Framework for the Analysis of Digital Communication

2.1 Introduction

In this chapter, we discuss the concept of genre as a theoretical framework for the analysis of the communication of knowledge on the internet, drawing attention to the complexity that digital media bring to this concept, as well as to the multiple affordances of digital genres for scholarly communication. First, a brief introduction of genre drawing on the Rhetorical Genre Studies (RGS) and the English for Specific Purposes (ESP) traditions is provided, focusing on those aspects of particular relevance to understanding how genres emerge, evolve and operate in the digital context. Then, we discuss some important issues and questions that the recent scholarly literature has raised in relation to digital genres, such as the relationship between the technical features of the medium and the genre in the digital environment, the evolution of genres through the incorporation of internet-enabled features and the difficulty of distinguishing between evolved/adapted genres and new genres. Finally, we examine the affordances of digital genres that support new ways of knowledge communication, including their potential for multimodal and multilingual communication. In closing this chapter, we summarize the framework that we advocate in the book to understand the notion of digital genre and to analyze the digital genres that scholars use.

2.2 Genre as a Theoretical Framework

2.2.1 What is genre?

Genre analysis is a theoretical and analytical approach based on a social view of language use in context. The concept of genre is versatile, and it has proved to be relevant across very heterogeneous fields of study. Within the broad field of applied linguistics, genre studies capitalize on the affordances of the concept of genre as a descriptive, interpretive and pedagogical tool. By way of illustration, a genre approach has been applied to describe professional communication practices (Bhatia, 2004; Orlikowski & Yates, 1994; Spinuzzi, 2003), to interpret social actions in

RGS (Bazerman, 1994; Miller, 1984) and to provide pedagogical orientations in English for academic purposes studies (Swales, 1990, 2004). For the purpose of this book, we approach genre from the perspective of applied linguistics, drawing on the ESP and the RGS traditions. The rationale for combining these two approaches is that both operate with a functional concept of genre, which is useful to explain why new digital genres for knowledge production and communication emerge, and what researchers do with them, as we will see in subsequent chapters.

Seminal definitions of genre in these two traditions provide clear insights into the meaning of this concept. Within the ESP tradition, Swales (1990: 58) defines genre as 'a class of communicative events, the members of which share some set of communicative purposes', the latter being recognized by the members of the discourse community that uses the genre. These purposes are manifest in linguistic patterns and genre conventions of form and content. Within the tradition of RGS, Miller (1984: 159) characterizes genres as 'typified rhetorical actions based in recurrent situations'. Genres are social actions that respond to an exigence, to an 'objectified social need' (Miller, 1984: 157), for example the need for self-expression, self-promotion and validation of one's own views in the case of a blog, or the need to democratize knowledge production and draw on the collective intelligence of the general public in the case of citizen science projects. Miller (2015b: 174) indicates that both the Swalesian concept of purpose and her own concept of exigence are concerned with the question 'Why?'. While 'purpose' addresses the question from the actor's point of view (i.e. 'Why are you doing this?'), 'exigence' addresses it from a system's point of view ('Why does this happen?', 'What does it achieve (...) for the stability of the system?') (Miller, 2015b). From this perspective, genres satisfy recurrent needs within a system. Bringing together both approaches enables us to better understand why digital genres for knowledge production and communication emerge and evolve (and decay), and why scholars use various semiotic resources to compose genres in the ways they do.[1]

In this book, we take up Miller's social and ecological view of genre, where genre is socially recognized and negotiated. We align with Miller's (1984: 163) contention that 'genre' is a construct that 'acquires meaning from situation and from the social context in which that situation arose', and as such, that genre is 'a rhetorical means for mediating private intentions and social exigence' (see also Miller, 2015a). The genres that we focus on in this book are tools for knowledge production and communication; these genres are thus forms of social action, used by researchers to construct their identity and enhance their visibility, to produce knowledge and to make the outcomes of scientific enquiry accessible and reachable to different stakeholders both within and beyond academic and research settings. Researchers today are developing new digital genres as typified responses to recurrent situations they encounter in their work. Users of these genres share knowledge about the communicative practices and conventions of the genres, which enables them not only to produce and interpret these genres, but also, as Freadman (2012: 547) claims, to adapt them to 'particular needs as occasions arise'. Miller (1984: 165) states that when we learn a genre we learn what we can do with it (e.g. apologize, account for progress in achieving goals), and thus genre knowledge helps us to understand how to participate in the activities and practices of a community.

The definitions of genre from the ESP and the RGS traditions consistently refer to sharedness, conventions, patterns, typification and recurrence. Therefore, the concept of genre implies similarities or regularities among texts that exemplify the genre. However, there is also variation among them, because every time a genre is used, it is performed in a unique way, to do a unique action in a unique moment (Devitt, 2015: 45). The tensions between similarity and uniqueness, between recurrence and contingency, result in the inherent dynamism of genres. Scholars in the RGS tradition have emphasized that genres evolve and change over time because when users perform a genre in a particular text, they do not need to draw on all the conventions of the genre (Orlikowski & Yates, 1994) and thus genres are often renegotiated. Dynamism and instability are inherent features of digital genres, due in part to the rapid technological changes, and also to the fact that many of these genres are less stabilized than are institutional genres of academic knowledge communication. They also evolve to respond to changing social needs (Mehlenbacher, 2019b). This dynamism has two important consequences for genre analysis. The first one is the need to focus on how digital genres emerge and are negotiated, and to analyze the influence of sociotechnical factors on the evolution of genres. The second is the difficulty of categorizing these genres and describing them as a stable class.

Another important consideration when discussing digital genres is that identifying them using the criteria of 'communicative goal or purpose' and 'intended audience' (Swales, 1990) seems to be a daunting and ineffective endeavor. Regarding the former criterion, Askehave and Swales (2001: 195) argue that although genre analytical methods have drawn on the criterion of the communicative goal for categorizing genres, this concept has become more complex. In digital environments, genres are usually hybrid in nature, serving a variety of interconnected purposes. Concurrently, the collapsing of contexts, or the flattening out of different audiences in the internet (boyd, 2002; Vitak, 2012), also makes the criterion of 'audience' problematic for genre identification because the traditional boundaries, for example, between expert-to-expert and lay communication are blurred in online environments. A productive concept for understanding communication in online settings is that of 'imagined audience', i.e. the producer's conceptualization of the collective he or she is communicating with (Marwick & boyd, 2011). For example, since social media offer few clues about the actual audience, writers often 'take cues from the social media environment to imagine the community' (Marwick & boyd, 2011: 115).

2.2.2 Genre relations in knowledge production and dissemination

Genres do not exist in isolation but interact with other genres in various ways. Spinuzzi (2004) uses the term 'genre assemblages' to refer to complexes of genres that are used in concert with each other, and discusses four frameworks to describe these assemblages: genre sets, genre systems, genre repertoires and genre ecologies. 'Genre repertoire' and 'genre ecology' are particularly useful concepts in the discussion of digital genres. The identification of 'genre repertoires', or the sets of genres that are deployed by a professional community (Orlikowski & Yates, 1994) to achieve all of such a community's purposes, can help us to understand the extent to which digital media have revolutionized academic communication. Internet affordances facilitate the emergence of new genres, thus contributing to the greater diversification of the repertoire of genres that have traditionally been deployed by researchers to disseminate new knowledge (Pérez-Llantada, 2021b; Swales, 2004). In addition to new genres for peer-to-peer communication, new genres for public outreach have also emerged as have genres blurring the boundary between expert and non-expert audiences. As an example, Kelly (now Mehlenbacher) and Miller (2016) use the term 'para-scientific genres' to refer to those genres that borrow authority and expert knowledge from conventional genres of internal communication, such as the peer-reviewed article, but operate outside the established norms of gatekeeping and reporting of these conventional genres. Blogs and crowdfunding proposals are examples of para-scientific genres. Later, Mehlenbacher (2019b) refers to these genres (e.g. science blogs) as 'transcientific genres', because they cross boundaries between internal and external spheres of discourse: although they are constructed by scientists and relate to other conventional scientific genres, they adopt strategies from external genres of science communication, such as popular science articles. A further important idea that Mehlenbacher underlines is the increasing use of online platforms that enable the lay public to have access to scholarly exchanges that have so far been restricted to the expert public.

The concept of 'genre ecology' is also particularly useful to explain how genres work together in the digital context. In the field of professional communication, genre ecology has been defined as referring to 'an interrelated group of genres used to jointly mediate the activities that allow people to accomplish complex objectives' (Spinuzzi & Zachry, 2000: 172). Spinuzzi and Zachry explain that within these genre assemblages 'content not only proliferates into different genres but also becomes interconnected in deliberate and persuasive ways'. The metaphor of 'genre ecology' in scholarly communication acts in the same way, for example, in digitally enhanced publications (Harmon, 2019; Pérez-Llantada, 2021b) or the research group blog (Luzón, 2017). Regarding the former, researchers can communicate their research through a traditional genre, the journal article, while at the same time enhancing and expanding the knowledge reported in the article by connecting this genre with other genres and multimodal elements (above all, through hyperlinking), such as hyperlinked-related publications, graphical abstracts, video abstracts or author summaries and podcasts. Wall and Spinuzzi (2018) claim that ecologies of genres can be strategically deployed to communicate disciplinary knowledge to wider and diversified audiences. For example, in the blogs written by research groups, genres are interconnected in opportunistic ways to accomplish complex objectives, among them to publicize a group's research to make it visible to the disciplinary community and connect with the interested public (Luzón, 2017). In genre ecologies, content is often reused or re-represented across different interrelated genres (Wall & Spinuzzi, 2018), and thus readers can select the specific content from the genre ecology as they wish.

Some genres for knowledge communication that are part of ecologies have been referred to as 'add-on genres' (Luzón & Pérez-Llantada, 2019) or 'appendant genres' (Yang, 2017). These genres have a relation of dependence regarding a host genre, without which they cannot exist (e.g. highlights, abstracts). According to Yang (2017), while the purpose of the host genre is usually to produce knowledge, that of the appendant genre is to share the information present in the host genre. For example, in response to the need to engage other scholars with one's research, the digital environment has facilitated the emergence of appendant genres that appear simultaneously with the research article (e.g. lay summary, video abstract), whose purpose is mainly to promote the article. The concept of 'add-on'/'appendant' genre is related to and overlaps with that of 'part-genre', i.e. a genre that is itself a member of a larger genre (Dudley-Evans, 2000) (e.g. article introductions, article highlights, thesis acknowledgements).

Two other types of relations between genres which help to explain how digital genres for knowledge communication operate are based on the concepts of 'uptake' (Freadman, 2002) and 'colony'. 'Uptake' refers to the dialogic relation between a genre and another genre that responds to it (Freadman, 2002). Many digital genres respond to content in a preceding genre in various ways (e.g. blogs discussing a research paper or assessing its reliability, open peer review reports). This relation, which in traditional genres often involves explicit reference to the uptaken text, is facilitated by the affordances of the internet (hypertextuality, modularity and interactivity) (see Section 2.3.3).

Finally, the concept of 'colony' refers to a group of 'closely related genres which to a large extent share their individual communicative purpose, although most of them will be different in a number of other respects, contexts of use and exploitation' (Bhatia, 2004: 66). As the members of a colony have a common purpose, they will also share some rhetorical strategies and discourse features. An example of a genre colony would be one made up by the genres used to assess research, discussed in Chapter 10.

In sum, assuming an ecological view of genre, with genres being dynamic, open, evolving and interconnected, leads us to expect that the changes in a genre may prompt changes in other genres to which it is related.

2.3 Digital Genres: Developing Languages of Description

From the early studies of digital genres it was clear that previous definitions of 'genre' were limited in that they did not consider the interplay between genre and medium, and that a reconceptualization of genre theory was necessary to account for these genres. In their pioneering study of genres on the internet, Shepherd and Watters (1998) define what they call 'cybergenres' in terms of the triple (content, form, functionality), thus adding functionality (i.e. the capabilities available in the new medium, such as hyperlinking or commenting) as a key element differentiating digital genres from pre-digital genres. Similarly, Askehave and Nielsen (2005) advocate an extension of Swales's (1990) definition of genre that incorporates media features, since, they argue, the medium affects the function, production and reception of digital genres. They propose a 'two-dimensional' model of analysis where genre is characterized both as text and as medium (with hypertextual or navigation possibilities) and claim that hyperlinking is an essential feature to define digital genres, since it influences how these genres are used.

Early studies classified digital genres by taking into account how they differed from pre-digital genres in terms of technical capabilities (i.e. functionality) (see Crowston & Williams, 2000). For example, in their analysis of the evolution of cybergenres, Shepherd and Watters (1998) consider the functionality afforded by the new medium as the main driving force. These authors classify digital genres into extant genres (based on genres on other media) and novel genres, depending on the level of functionality added by the medium. Extant genres can be replicated (i.e. faithful reproductions, both in content and form, of genres existing in other media with little functionality) or variant (i.e. genres that have evolved by adding new functionality). Novel genres are subdivided into emergent (i.e. derived from existing genres but incorporating functionalities that make them significantly different) and spontaneous genres (genres with no clear counterpart in other media, such as the hotlist, i.e. a list of hyperlinks to websites on a topic).

With the rapid development of the Web 2.0, the explosion of digital genres in the last two decades and their fast evolution, and the high level of experimentation facilitated by internet technologies, a redefinition of the concept of (digital) genre was necessary. Bateman (2008: 212) suggests that characterizing digital genres in terms of functionality attributes (e.g. browsing, email, search, interaction) is questionable, since these attributes are very general, occurring across digital genres and not distinctive of specific genres. Another important limitation of previous definitions of (digital) genres is that they have neglected their inherently multimodal nature (Bateman, 2008, 2014; Herring, 2019). Kress (2003) posits that one of the problems in adapting genre theory to new media is that much genre theory has been developed for the analysis of linguistic monomodal texts, rather than multimodal artifacts. Similarly, Herring (2019) argues that the analysis of computer-mediated communication (CMC) requires moving from linguistic discourse analysis to multimodal discourse analysis.

Miller and Fahnestock (2013) also call for more conceptual clarity to answer several broad questions about digital genres for knowledge production and communication:

> Can genres be said to evolve? Do they evolve from other genres? From adaptation to changing rhetorical situations? (…) what [is] the difference between an affordance or medium or mode of communication and a genre? How are new genres identified? What brings them about? (Miller & Fahnestock, 2013: 3)

These questions are pertinent to the study of digital genres in this book and thus we approach them below by briefly discussing the criteria that have been used to define digital genres, and the evolution of these genres and their affordances.

2.3.1 Defining digital genres: Form and function, medium and mode

An important issue in the definition of digital genres is the form/function distinction: what is relevant when defining genres – form (e.g. linguistic and rhetorical features), function (e.g. communicative purpose) or a combination of both? (Heyd, 2016). Although early definitions of genre were mostly form centered, since Miller's (1984) definition of genre as social action, a pragmatic approach has prevailed. More recently, researchers on digital genres have stressed the need to consider both form and function when analyzing these genres, although they add that in the digital environment a more complex, dynamic and rhetorical view of form is necessary (Devitt, 2009; Puschmann, 2009) and thus 'form' has been extended to include technical features (Heyd, 2016; Puschmann, 2009). For instance, among the formal features of blogs, Puschmann

(2009) includes the segmentation of text into entries, the archive or the blogroll (list of blogs, usually placed in the sidebar, that the blogger wants to share).

The complex relation between technical features afforded by the medium and genre in the digital environment has been discussed by several scholars who have often used the blog as an example. Lüders *et al.* (2010), for instance, reject the claim that medium is an integral part of the genre and argue that they should be kept separate. They characterize the blog as a medium (based on technical features such as the use of specific templates, comments, presentation in reverse chronological order), and consider that different blog genres are defined in terms of the conventions they apply and the expectations they meet, as a result of their use in specific social contexts. For other authors, the form/function distinction (with form including technical features) helps to group together genres with 'family resemblance' (Wittgenstein, 1953) and establish hierarchies of supergenres (i.e. broad categories) and subgenres. For instance, Puschmann (2009) sees the blog as a supergenre, with subgenres such as campaign blogs, health blogs or news blogs, among others. All these blogs share some formal technical features (e.g. entries, reverse chronological order, use of hyperlinks) and some formal linguistic features (e.g. frequency of first-person pronouns), but the function is different among subgenres, because they respond to different exigences, i.e. different 'objectified social needs' (Miller, 1984).

Puschmann (2009) and Lomborg (2011) define the supergenre (e.g. the blog) in terms of formal criteria and technical affordances, while the subgenres (e.g. types of blogs) are functional categories. Heyd (2008), by contrast, introduces a two-level genre model where the supergenre is defined functionally and the subgenres are defined formally (see Figure 2.1). She posits that all these subgenres have a common genre antecedent and share a relatively stable communicative function, but are formally different because of genre evolution and adaptation to the medium. This view, however, suggests that genres evolve and emerge because of the existence of technical capabilities; instead, we believe that new genres emerge or develop to respond to 'objectified social needs' that function as rhetorical motives (Miller, 1984). The research group blog, for instance, has emerged because of the underlying exigence of the self-promotion of research groups.

When analyzing digital genres, analysts can focus on supergenres and adopt broad general labels (e.g. blog, homepage), or on subgenres (e.g. corporate blogs, academic department homepage). In this book, we adopt a fine-grained approach in our analysis and labeling of digital genres, and focus on subgenres (e.g. research group blogs), since, following RGS, we view genres for knowledge production and communication as socially situated and responding to particular rhetorical exigences.

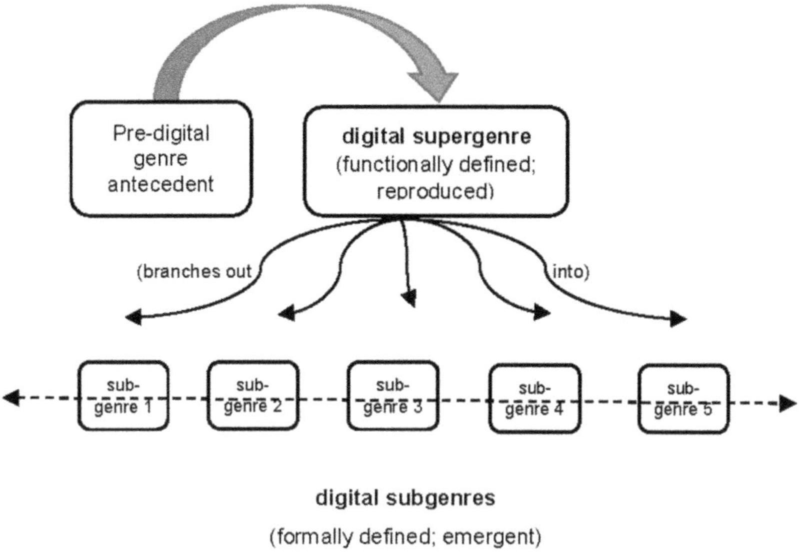

Figure 2.1 A genre model of digital supergenres and subgenres (adapted from Heyd, 2008: 201)

The discussion of form and function above has drawn attention to the formal technical features of genres afforded by the medium, but not to the inherently multimodal nature of digital genres. We align with Herring's (2019) claim that CMC, and therefore digital genres, should be reconceptualized as 'fundamentally multimodal'. Text is only one possible semiotic resource, and most digital genres include a variety of modes and resources (e.g. writing, sound as speech, music, still and moving images). When composing digital genres, users select semiotic resources from those afforded by the medium and orchestrate them in such a way that they work together to realize specific rhetorical strategies and thus achieve the communicative purpose(s) of the genre (Bateman, 2008, 2014; O'Halloran, 2009).

2.3.2 Evolution and remediation of genres

The migration of traditional genres to the web and their evolution in the digital medium has been a dominant line of research in relation to digital genres. Distinguishing between reproduced/replicated genres, with their variants, and emergent genres is a challenging endeavor, since, as Rowley-Jolivet and Campagna (2011: 45–46) state, digital genres can be placed on a cline from replicated genres, which 'have moved intact to the web' to new web-native genres (e.g. the homepage). Along the same lines, Herring (2013) notes that over time web genres tend to move along a continuum of adaptation, evolving towards 'emergence'. She illustrates this claim with the social network site Facebook, which had as an antecedent print-format 'face books' from Harvard University dormitories but has added and combined many internet-enabled features such that 'the whole is qualitatively different from the sum of its parts, and it has no single offline or online precedent' (Herring, 2013: 7). When focusing on genres for knowledge communication, we also find examples of this process with the migration of genres such as the research article or the laboratory notebook to the web. The research article in PDF format is an example of a replicated genre, whereas enhanced research articles, which still respond to the rhetorical exigence of printed articles, add internet-enabled features (e.g. interactive visual elements and tools, as well as multimedia), thus falling in the middle of the cline (Pérez-Llantada, 2013). Further along the cline, we find other publication formats, such as the 'Rich Internet Publication' (RIP), which, according to Breure *et al.* (2011), cannot be described as 'enhanced' publications, since they are 'high quality multimedia presentations', more similar to web applications than to the printed journal article. At the emergent pole of the cline, we could also place new forms for reporting research results, such as databases (Crowston & Williams, 2000) or new instantiations of genres in the form of applications for mobile web use (Heyd, 2016).

The adaptation of non-digital genres to the medium cannot be understood without reference to the concept of remediation, that is, 'the formal logic by which new media refashion prior media forms' (Bolter & Grusin, 1999: 273). Many digital genres are constructed through a process of remediation, in that traditional genres are imported into new media where they evolve into variants afforded by the technical capabilities of the new medium (Bawarshi & Reiff, 2010). In this process, genres are recontextualized, transformed and repurposed to fit a new context. One interesting question for the definition of genre is posed by Giltrow and Stein (2009: 9): 'Does a new medium automatically make for a new genre? Is it possible for a traditional genre in the spoken or written media to migrate into the Internet without loss of identity?'. Heyd (2016) uses the academic article as an example of a genre that has not experienced much change after migrating to the web, a claim also supported by Mackenzie Owen (2007) and Gross and Harmon (2016: 50), who contend that the web has not changed the essence of the scientific article. Interestingly, Mehlenbacher and Mehlenbacher (2019: 47) suggest that this might be so because the online article in its current form responds to the same rhetorical exigence as the printed article: 'to share research findings with experts'. This supports the view that new genres appear because there is a latent social need. In the case of the scientific article, even if the nature and format of this genre have not changed much on the internet, the need for the reproducibility of research has triggered the emergence of new genres which provide a detailed record of methods, such as the Registered Report, a type of article where the methods are reviewed and accepted before data collection begins (Mehlenbacher, 2019a), or the video methods article (Hafner, 2018) (see Chapter 9).

Other genres for knowledge production and communication have undergone more radical changes than the research article to meet new rhetorical needs. This is the case of the lab notebook, which has developed into the electronic notebook and further into the open lab notebook to facilitate the sharing of research protocols and collaboration with other researchers (Wickman, 2016). Another example is the crowdfunding proposal (Mehlenbacher, 2017), whose antecedent is the conventional research funding proposal. When compared with the traditional proposal, the crowdfunding proposal is composed to appeal to a wider audience and persuade it to fund the research. It therefore employs a variety of multimodal resources, such as videos or images, not used in the traditional genre.

This process of remediation and adaptation by incorporating internet-enabled features has therefore shaped the generic features of digital genres, often giving way to hybrid genres with features and elements from different genres. A clear example of the hybrid nature of digital genres is the blog (Luzón, 2017; Mauranen, 2013). Herring *et al.* (2004) view the blog not as a new genre, but as a hybrid of existing genres, sharing features with personal homepages and asynchronous discussion forums. In her study of research group blogs, Luzón (2017) shows that these blogs may incorporate a variety of genres through embedding (e.g. abstracts, biographical notes) or linking (e.g. homepages, articles). Another example of hybridity is the video method article (Hafner, 2018), where the multimodality afforded by the video makes it possible to combine features of a conference presentation, such as direct audience engagement, with the demonstration of laboratory methods.

2.3.3 Affordances and features of digital genres

The concept of affordance, coined by the psychologist Gibson (1966), pervades the literature on digital genres, as it helps to explain the potential for action and for meaning construction that these genres offer. Hutchby (2001: 444) defines affordances as 'functional and relational aspects which frame, *while not determining*, the possibilities for agentic action in relation to an object' (our emphasis). This definition emphasizes the agency of the user, who decides whether and how to use the potential of the object (e.g. technologies) to perform specific actions. Miller and Shepherd (2009) agree, stating that the affordances of the internet medium do not determine the adaptation of traditional genres to the online medium or the emergence of new genres; rather they interact with 'an exigence that had not been yet met', they coax 'into being a latent social motivation that, when available, is instantly recognizable to large numbers of people' (Miller & Shepherd, 2009: 282).

Jones and Hafner (2012: 5–9) categorize the affordances of media into five types that can affect:

(1) the kind of actions that people can perform (e.g. sharing photos and information);
(2) the kind of meanings that people can express (e.g. meaning can be expressed visually on multimodal web pages);
(3) the social relations people can have (e.g. in blogs people can comment and interact);
(4) the ways people can think; and
(5) the social identity that people can assume (e.g. digital media enable users to present selective information about themselves).

Other researchers (Baym, 2010; boyd, 2010; Hopkins, 2015) mention these affordances: storage and searchability (the ability to find content easily through a search), modularity (use of 'modules' to compose digital documents, which facilitates the integration of modes in a single platform), wide reach and increased visibility of content, interactivity, multimedia and hyperlinking. In the following, we briefly discuss some affordances we feel are particularly important in shaping digital genres and supporting new ways of knowledge production and communication.

2.3.3.1 Hypertext and modularity: Linking and embedding

Two affordances which have profound effects on how digital genres are constructed and which result in complex intertextual relations are the ease of reusing and embedding existing texts and the possibility of connecting to other texts by linking. Digital genres are often composed by incorporating other genres or linking to them to form complex documents that can serve multiple purposes. Genre embedding, or the incorporation of a text representing a genre within the matrix of another distinct genre, is a common practice in digital genres (Kwasnik & Crowston, 2005), facilitated by the technological affordances and the flexibility of composing in this medium. For instance, a video abstract can be embedded on a research group's website or blog.

Hypertext is also a fundamental feature of digital genres, providing a tool for writers to organize information within a single website (through internal links) and to include prior texts within their text and make them available to readers (Jones & Hafner, 2012: 36). Text producers can use links to establish semantic relations between the texts that are linked and create contingent relations between genres. In addition, hypertext allows readers to engage in more active reading by choosing their reading path. Askehave and Nielsen (2005) argue that when reading digital texts, readers become navigators who decide how to construct the text that they are reading, i.e. whether to read linked texts and in which order to read them.

The linking and composing affordances of digital genres shape them as open, fluid and hybrid, and facilitate the creation of relations between genres not existing in other media. Catenaccio (2012: 40) sees websites as 'hypertextually organized rhetorical interface[s]' and emphasizes that while they are self-contained even if porous and flexible entities, they embed multiple genres. These genres are part of an ecology, connected by the producer in deliberate and opportunistic ways, which work together to achieve the overarching purpose of the website. The hypertextual nature of websites facilitates the combination and mixture of elements in other media and modalities.

Hyperlinking and modularity also enable interaction between genres that may require rethinking previous models of genre relations and intertextuality. These affordances create close relations between hyperlinked genres (e.g. a scholar's homepage and his/her online research papers) and between 'neighbouring genres' (Heyd, 2016: 96) which coexist in the same space, such as a Wikipedia article and Wikipedia talk pages. Casper (2016) notes that none of the previous models of genre relations can satisfactorily account for the relations of 'simultaneous presence' established through linking between some hyperlinked genres. Indeed, digital genres are not just related intertextually, 'they are linked materially' (Casper, 2009: 148).

2.3.3.2 Media convergence and multimodality

The technological capabilities of the medium enable the convergence of many technologies – e.g. text, voice, images, computing – and the mixing of a variety of modes in a single platform to create 'hybrid' multimodal genres (Adami, 2014; Askehave & Nielsen, 2005; Heyd, 2016). Multimodality is not exclusive to CMC, but the possibilities for mode combination afforded by the web have led to radical changes in the production and interpretation of print genres that have migrated to the internet. A good example is the research article. When published online, the print research paper can be enhanced with visual elements such as videos, interactive maps, animation and audio files (Aalbersberg *et al.*, 2014; Breure *et al.*, 2011). Through visualization, digital publications may provide much more detailed documentary evidence to support arguments than do print publications (Jakubowicz & van Leeuwen, 2010). The multimodal nature of some online papers facilitates new ways not only of representing content but also of interacting with content, e.g. readers can watch videos, interact with graphs, tables and 3D structures, and access related material (Gross & Harmon, 2016).

Another type of digital publishing format with which some humanities journals have experimented is the webtext. Webtexts are texts designed with elements and media appropriate to the web (e.g. links, java script, streaming media and animations) and therefore they would lose the power of argument if they were printed (Ball, 2016). *Kairos: Rhetoric, Technology, and Pedagogy* (see http://kairos.technorhetoric.net) was the first journal to exclusively publish webtexts. In its 'About' page, the editors state that the journal promotes work that constructs academic argument through the use of new media. The *Journal for Artistic Research* also provides a digital platform to allow artistic researchers to publish their work (https://www.jar-online.net/journal-artistic-research). This self-acknowledged purpose provides evidence for Cope and Kalantzis's (2014) statement that digital publications enable researchers in disciplines such as the arts, media and design to participate in the academic knowledge system with outputs other than the traditional article.

Studies on multimodality and multimodal genres have emphasized that different modes have different potentials for constructing meaning ('modal affordances') and therefore each genre, depending on the modes that it integrates, also has specific semiotic affordances (Jewitt, 2016; Kress *et al.*, 2001). In the context of knowledge communication online, an increasing number of studies are exploring how multimodality is being harnessed in a wide range of genres (e.g. blogs, videos, podcasts, enhanced papers, journal websites) to convey meaning and represent it in more effective and engaging ways (Engberg & Maier, 2015; Hafner, 2018; Rowley-Jolivet & Carter-Thomas, 2019). The convergence of media afforded by the web has also enabled the appearance of complex hypermodal (i.e. conflating multimodality and hypertextuality) websites that facilitate knowledge production and communication. An example is the hypermodal environment in which research video articles from the *Journal of Visual Experiments* (JoVE) are embedded. Engberg and Maier (2019) show how the genres embedded in this environment (the video, the research article, downloadable documents) are interrelated in such a way that knowledge building is enhanced.

2.3.3.3 Interactivity

The interactivity embedded in Web 2.0 enables users to interact with the text and with each other in ways that are not possible in other media (e.g. users may comment on blogs, respond to a YouTube video, edit Wikipedia pages). In the context of knowledge production and communication, this affordance supports public engagement with science (Brossard & Scheufele, 2013) to a great extent and has also enabled new writing and reading practices. Scholars can comment on various digital spaces (e.g. open lab notebooks, science blogs) and thus provide feedback, and they can share different types of resources and participate in idea-generating interactions or engage in discussions. The interested public can easily search for information and can participate actively in science-related discussions or even contribute to scientific research (e.g. citizen science), thus widening the science participatory framework.

The public can also interact with online information by sharing it (e.g. sharing science videos on YouTube or retweeting scientific tweets), thus blurring the line between users and providers of content, or by 'liking' it (Brossard, 2013; Kahle *et al.*, 2016). Researchers can provide post-publication feedback and reviews of articles published in some online journals (Casper, 2009). In addition, the possibilities for interaction enabled by social media also lead to the formation of online communities and social networks that facilitate informal knowledge production and sharing.

2.3.3.4 The language affordance: Digital genres and multilingualism

Since the publication of *The Multilingual Internet* (Danet & Herring, 2007), research on multilingualism online has increased, 'multilingualism' here referring to various linguistic phenomena. Barton and Lee (2017: 142) distinguish two types of studies within the area of multilingualism online. The first type of study has explored linguistic diversity online, focusing mainly on the distribution of English and other languages, and the number of languages available and used in particular digital spaces and platforms. Although the internet is a multilingual space, English is still the language with the highest presence on the internet, closely followed by Chinese (Internet World Statistics, 2019); and, what is more relevant for this book, it is often the lingua franca online (Tagg, 2015; Yu *et al.*, 2018), particularly for knowledge communication and dissemination. This dominant role of English allows researchers from different linguacultures to communicate and collaborate across linguistic and geographic boundaries; however, it also raises the question of up to what point digital genres facilitate the participation of non-Anglophone scholars in global academic discussions and knowledge production, and has important consequences regarding who is heard and seen or who decides what topics are worth discussing.

The second stream of research on online multilingualism is concerned with language practices online and online interactions among multilingual users. Researchers have explored phenomena such as code-switching and language choice online (Androutsopoulos, 2007; Seargeant *et al.*, 2012) or the use of multilingual resources to perform particular identities (Lee, 2016). Several studies have approached multilingual practices on the internet from the perspective of translanguaging (e.g. Han, 2019; Kulavuz-Onal & Vásquez, 2018) and thus described 'the ways in which groups and communities of people experience and do things that involve more than one language' (Barton & Lee, 2013: 60–61). This broad definition of translanguaging helps to account for the complexity and variety of linguistic practices online. These practices usually involve integrating various linguistic and semiotic resources and can be engaged in by users who have resources from more than one language, even if they are not proficient in those languages (Kulavuz-Onal & Vásquez, 2018).

Multilingual practices in digital contexts are influenced by the fact that producers are 'digitally connected to other individuals and groups' (Androutsopoulos, 2015: 188). Research on knowledge communication online has revealed that some digital genres are intended to reach diversified networked audiences and this determines the producers' multilingual practices (e.g. Luzón, 2013). In addition, the internet enables integrating content in several languages in a single platform and linking between documents with different languages, thus creating multilingual texts. Since genres on the web are interconnected, one genre in a given language may interrelate through the web's hypertextual affordances with other genres in different languages, by this means supporting multilingual practices (Luzón, 2017; Pérez-Llantada, 2021a, 2021b).

2.4 Concluding Remarks: 'Genre' in this Book

In this chapter, we have presented our own conceptualization of 'genre' and of 'digital genre' in particular. We have explained why we view genres as forms of *social action*, as tools to achieve *particular purposes* by using various *semiotic resources* strategically. We have also explained that we align with the view of genres as *typified responses* to recurrent rhetorical and social situations, which makes it necessary to analyze genres as situated within social contexts, constructed for particular purposes, and as discourse types that share regularities in terms of content, rhetorical structure or linguistic patterns. We have also noted that while these regularities are motivated by and reflect the communicative purposes that the genre is intended to achieve in the particular social context, there is variation in performance between instances of a genre, since writers do not always adhere to all the conventions of the genre.

We have also explained that digital genres are inherently *multimodal* and *hybrid*, including elements from various genres. They are related to many other genres in complex ways, and they tend to be highly dynamic and, for this reason, digital genres cannot be understood if studied in isolation but rather in close relation with other genres. As stated previously, genres form different types of *assemblages* with other genres, working together with them to accomplish actions; they often respond to other genres in various ways; and they may be related to antecedent genres, through a process of adaptation and evolution. We have also brought to the fore the view of digital genres as being in constant flux, mainly because of rapid changes in the technological capabilities of the internet, which makes it possible for researchers to adapt the genre to respond more effectively to new demands. Therefore, as we will also illustrate in the following chapters, the concepts of innovation and creativity are basic to understand digital genres. Since digital genres are open, evolving classes, categorizing genres and establishing the boundaries between a digital genre and its (printed or digital) antecedent(s) is often a daunting enterprise. Moving to the digital medium does not automatically turn a traditional genre into a new genre (e.g. the research article), but digital media offer new possibilities to achieve new purposes, which drive the evolution of the genre.

The concept of genre presented here thus guides genre analysis in this book. Conceptualizing genres as tools for accomplishing social actions and achieving communicative purposes implies that the main goal of genre analysis is to find out why genres emerge, what users do with them and why they are composed the way they are. The study of digital genres for knowledge dissemination therefore involves analyzing, among other aspects, the *rhetorical exigences* to which these genres respond, the *audiences* at whom they are addressed (e.g. their level of expertise), *their communicative purposes* and the *rhetorical strategies* to achieve these purposes, the various *modes* (written, oral, visual) made available in the genre and how they are combined to realize rhetorical strategies and make meaning, the different *types of discourses* that they integrate, the *affordances* and *constraints* of the medium or the *types of interaction* they support (e.g. readers' responses, editing by other users). When dealing with digital genres, analysis has to be adapted taking into account their distinctive features, such as *hybridity*, *modularity*, *hypertext* and *multimodality*. Given the multimodal nature of many digital genres, it is necessary to adopt a semiotic approach when analyzing them (see Bateman, 2014; Kress, 2003; van Leeuwen, 2005), which means that the focus of analysis should not only be the written text but all the semiotic resources that contribute to achieving the communicative purpose of a particular genre.

The concepts introduced in this chapter will be explored in further detail and applied in the forthcoming chapters, in which we discuss why digital genres emerge, what researchers do with them and how these genres are changing the dynamics of knowledge production and communication in the current research and academic context.

Note

(1) The value of combining rhetorical and linguistic approaches has been pointed out by other researchers (Devitt, 2015; Mehlenbacher, 2019b).

3 Knowledge Communication in the Digital Era

3.1 Introduction

Some of the most far-reaching changes in scholarly communication have resulted from the use of the internet, which has transformed how researchers share information and communicate the outcomes of their work in two fundamental ways. First, the functionalities of the web allow researchers to produce new knowledge, share it and disseminate it widely and openly in a much faster and more flexible way than ever before. Secondly, Web 2.0 offers a wide repertoire of emerging genres for communicating with both the expert and the non-expert public, through these means supporting new practices for knowledge communication that go beyond peer-to-peer knowledge exchange.

Nonetheless, technological development is not the only factor affecting knowledge production and communication in the 21st century. Socioeconomic demands and interests together with cultural and linguistic factors are also driving forces in how research is conducted and communicated in this digital era, and therefore shape the genres that scholars use. Thus, in order to contextualize our analysis of digital genres for knowledge communication, in this chapter we discuss the socioeconomic context within which researchers work, the material, technological and linguistic affordances and constraints that shape researchers' activity, and how digital media influence knowledge communication today.

3.2 The Changing Context for Knowledge Production and Communication

In the current competitive research context, many researchers need to meet the demands of multiple stakeholders, as the allocation of funding depends largely on the impact of their research output on knowledge, society and economy. This new approach to knowledge production and communication reflects broader social and socioeconomic dynamics that have fueled increasing multilocational (cross-national) cooperation and cooperative relationships between social agents and states (see Santos, 2002, for further discussion). In the context of the increasing marketization of knowledge produced in higher education institutions, and the view of universities as manufacturers of knowledge (Knorr-Cetina, 1981; Pérez-Llantada, 2012), research output stands as a highly valued asset for knowledge-intensive economies. It is therefore not surprising to see how both the existing research agendas and evaluation systems that assess the quality of knowledge production remain aligned to models in which academic productivity and research products have gained the status of 'commodities' (Lillis & Curry, 2010; Pérez-Llantada, 2012; Santos, 2002). In this market-driven environment, impact, visibility, collaboration and efficiency are, doubtless, key targets for scholars, which is reflected in their communication practices.

A consequence of these dynamics of knowledge production in a highly competitive research world is the pressure on researchers to publish in high impact journals in order to achieve recognition and advance their careers (Lillis & Curry, 2010; Pérez-Llantada, 2012; Swales, 2004). Research and publication in these journals tend to be the main criteria according to which scholarly success is measured and are therefore key in promotion and tenure processes and in enhancing scholars' reputation and recognition from the international community (Acord & Harley, 2012; Herman & Nicholas, 2019). Hence, researchers' increasingly need, or wish, not only to publish in high-ranking journals but also to give visibility to their research output. Visibility contributes to achieving reputation and prestige and brings benefits for the scholars or research groups who possess it (e.g. invitations to participate in research collaboration, funding for research). In order to increase the academic, economic and societal impact of their research, many scholars are expected to make it visible not only to international expert audiences, but also to various types of national and local audiences, which may entail the use of English (Corcoran *et al.*, 2019; Kuteeva & Mauranen, 2014) and the use of local languages, including for public outreach (McGrath, 2016). Digital genres can be effective tools to provide disciplinary, transdisciplinary and public visibility (Herman & Nicholas, 2019), as they make it easier for various audiences to discover, access, discuss and share research outcomes.

In this context, collaboration and networking are invaluable tools for increasing research impact and academic visibility (Curry & Lillis, 2010). The fact that knowledge production has become measurable and citations have become important may explain the increasing research networking activity witnessed across disciplinary knowledge domains. Using bibliometric analysis based on Thomson Reuters' Web of Science Citation Index Expanded, Soete *et al.* (2015: 36) identify an upward trend in publications with multiple authors from different countries, which reflects the importance of cross-border collaboration. In digitally networked environments, which facilitate knowledge networks, research collaboration and co-authorship between authors from different geopolitical regions are likely to continue increasing in the future. Interdisciplinary collaboration is also an important trend in the contemporary research context, and an area of concern and interest for funding agencies and research-intensive institutions (Wernli & Darbellay, 2016). Interdisciplinary collaboration is particularly valuable and necessary for addressing large-scale complex problems, including climate change, economic inequality and education (Demarest & Sugimoto, 2014). Technological improvements make it possible for researchers to access huge amounts of data in different disciplines and to communicate with other researchers, which facilitates working across disciplines (Wernli & Darbellay, 2016).

Another demand for researchers in many disciplines, arising from the need to conduct cost-effective research and to make efficient use of resources, is that research outcomes should be reproducible and reusable. This social exigence results from the 'science reproducibility crisis', that is, the fact that the experimental results of some studies cannot be replicated due to a lack of transparency (Baker, 2016), which raises questions about their validity. The demand for transparency in the research and publication process has also resulted in the emergence of various practices to support transparency in academic assessment (see Chapter 10), including new open peer review models, as in *Frontiers* (2016), which offers an interactive and collaborative review to guarantee quality standards, and online platforms such as *Publons*, which give credit to reviewers' unpaid work. Transparency in the whole research cycle is one of the goals of the open science movement, which we discuss below. This movement reflects a new research culture, which promotes openness, sharing of data and reproducibility. Over the last few decades, the model of research assessment based on productivity has put pressure on researchers to compete for the publication of results. However, society and many research-funding institutions are increasingly encouraging researchers to adopt an open science model (i.e. to engage in open collaboration rather than competition) as the best way to maximize the impact of their research. This tension between competition and collaboration reflects the conflicting demands that many scholars need to balance in the current complex research context.

Before discussing open science, it is important to note that not all researchers across geolinguistic contexts and academic disciplines need to meet the same demands nor do they work within the same linguistic and socioeconomic context to meet them. Studies focusing on the geopolitics of academic writing use the terms 'center' (or 'core'), 'periphery' and 'semiperiphery' of knowledge production to highlight the inequalities in academic publishing (Bennett, 2014; Canagarajah, 2002; Lillis & Curry, 2010). 'Center' is used to refer to Anglophone countries and other countries in Western Europe, with ample resources to publish internationally (Bennett, 2014). The boundary between 'periphery' and 'semiperiphery' is not clear-cut, these terms being used to refer to countries whose researchers are disadvantaged (linguistically or economically) vis-à-vis researchers from the center. Peripherality in academia is not, however, exclusively determined by geolinguistic location. Not all the scholars in the Anglophone center work in well-resourced contexts (Swales, 2019a), and there are also differences in semiperipheral countries, such as Spain, depending on the disciplines, with scholars from the hard sciences having more funding available for research than those from the soft sciences (Ferguson *et al.*, 2011). In this book, we use the term 'periphery' to draw attention to the challenges faced by some scholars as a result of the uneven availability of linguistic and material resources to conduct and publish research.

3.3 Open Science: Demands for Participation and Transparency

In recent decades, we have been witnessing a blurring of boundaries between scholarly discourse and public communication of science (Puschmann, 2015a; Trench, 2008). This blurring of boundaries results not only from technological innovations that enable scholars to communicate directly with the public, but also from a more profound sociological change: the increasing democratization of knowledge and the demand for a more participatory framework of knowledge production and communication (Puschmann, 2015a; Wilkins, 2008).

This push for the democratization of knowledge has led to a paradigm shift in some models of research and in the communication of research results: open science. Digital genres go hand in hand with open science, to the point that the Organization for Economic Co-operation and Development (OECD, 2015: 7) defines open science as making 'the primary outputs of publicly funded research results – publications and the research data – publicly accessible *in digital format* with no or minimal restriction', although open science is not in fact restricted to open access, but to the openness of the whole research process. Indeed, Dallmeier-Tiessen and Simko (2019: online) include 'open policies' within the concept of open science, which they define as encompassing 'all aspects of how scientific research is governed, performed, shared, published and evaluated'. It should be noted, however, that although knowledge communication online is not always open, the internet makes it easier to share ideas, data and results, thus enabling open science. The report 'The Rationales of Open Science: Digitalisation and Democratisation in Research', published by Science Europe (2017),[1] states that digitalization contributes to the openness of the science system by facilitating access to scientific information, making data available and reusable, facilitating collaboration and multidisciplinary research and providing tools to engage citizens. As many digital genres for knowledge communication have arisen in response to the need for making science more accessible and knowledge production more participatory at all stages of research, it is useful to consider the features of open science in order to understand the rhetorical exigences driving the emergence of these genres.

According to Vicente-Sáez and Martínez-Fuentes (2018: 428), 'Open science is *transparent* and *accessible knowledge* that is *shared* and *developed through collaborative networks*' (emphasis added).

(1) *Transparency* relates to the idea of opening up research and representing it in a way that enables reproducibility, peer control and reusability by other researchers (Hampton *et al.*, 2015; Rentier, 2016).
(2) *Accessibility* involves making data and scientific ideas available (online) to a broad population and making the products of research freely accessible (or at a very low cost) to everyone (European Commission, 2015; Hampton *et al.*, 2015).
(3) *Sharedness* puts the emphasis on sharing the whole research process: ideas, data, research protocols, methods, new findings.
(4) *Collaboratively developed knowledge*. Attention is drawn to the importance of Web 2.0, web-based tools and advanced technologies to facilitate cooperative work and global collaboration, and to the need to collaborate at different stages of research and across disciplines and roles (European Commission, 2015; Hormia-Poutanen & Forsström, 2016).

The features of open science identified in the literature by Vicente-Sáez and Martínez-Fuentes (2018) relate to the five open science schools of thought distinguished by Fecher and Friesike (2014), whose assumptions are presented in Table 3.1.

Open science, as conceptualized by these schools of thought, both accounts for and is facilitated by the emergence of new digital genres for knowledge communication. The *democratic school* is concerned with the concept of open access, that is, free access to the products of research (e.g. data and source materials, research publications) (Fecher & Friesike, 2014). The main argument is that the scientific outputs of

Table 3.1 Five open science schools of thought

School of thought	Central goal	Tools and methods
Democratic	Making knowledge freely available for everyone	Open access, intellectual property rights, open data, open code
Pragmatic	Opening up the process of knowledge creation	Wisdom of the crowds, network effects, open data, open code
Public	Making science accessible for citizens	Citizen science, science public relations, science blogging
Infrastructure	Creating openly available platforms, tools and services for scientists	Collaboration platforms and tools
Measurement	Developing an alternative metric system for scientific impact	Altmetrics, peer review, citation, impact factors

Source: Adapted from Fecher and Friesike (2014: 20).

research (especially publicly funded research) are a public good and therefore everyone should have equal right to access them. Open access facilitates reproducibility, efficiency and good use of public funding (Sarcina, 2019), and improves the quality assurance system of science and its credibility (Science Europe, 2017).

The affordances of the internet and digital genres facilitate not only open access to research data and output by researchers from all over the world (as we will discuss in Chapter 5 with the case of open laboratory notebooks), but also the interconnection between open publications and open data that is necessary for true open access (Boulton, 2012; Sitek & Bertelmann, 2014). In addition, these new media affordances enable researchers to present information in a way that facilitates reproducibility and reuse (see discussion of video method articles in Chapter 9). Of course, knowledge is not always communicated openly, since many digital texts (particularly in some publisher-based platforms or journals) are not freely accessible. That is why open access through online platforms is particularly important for knowledge production, especially in countries where researchers lack funds to access articles behind paywalls, and for the dissemination and sharing of research produced in the periphery (see Alperin *et al.*, 2008, for a discussion of the wide adoption of open science in Latin America; Salager-Meyer, 2018; Sarcina, 2019). An interesting initiative is the Scholarly Communication in Africa Programme (SCAP) (2010–2014), based on the assumption that openness (open access, collaborative research, digitally mediated sharing of resources) is crucial to allow African researchers to participate in global knowledge production and for their research to reach the audiences that can best exploit it for local development (Trotter *et al.*, 2014). It should also be noted that the adoption of an open access model by journals may not only facilitate access to publications by scholars in the periphery, but it may also perpetuate inequalities in knowledge production. It may make it difficult for some of these scholars to publish in high-impact open access journals that, in order to compensate for the loss of subscription revenue, have adopted a pay-to-publish model, and have high article processing charges (APCs), which neither the authors nor their institutions can pay (Demeter & Istratii, 2020). In addition, this pay-to-publish model of open access has contributed to the development of 'predatory open-access journals', that is, sub-standard journals that exploit this model for their own benefit. The pressure to publish in English may make some novice and periphery scholars prey to these predatory journals (Fazel & Heng Hartse, 2018).

Open access, not only to output but also to data protocols, is also essential for the *pragmatic school*, which is concerned with collaboration and building on the work of others in order to make the knowledge construction process more efficient (Fecher & Friesike, 2014). Nielsen (2012) emphasizes the importance for collaborative research of online tools that enable different stakeholders (including non-experts) to participate in the research process. Collaboration requires transparency in the process of knowledge creation. Open lab notebooks are an example of digital genres that increase the transparency of experimental methodology and thus facilitate collaboration (Nielsen, 2012). The recent Covid-19 pandemic provides a clear example of the importance of open access for collaboration to solve global problems. Researchers, journals (including *Nature*, *Cell* and *The New England Journal of Medicine*, all normally behind paywalls) and scientific institutions (such as the British Academy of Medical Sciences) shared their data and findings over the internet in unprecedented ways to collaborate in finding a way to prevent or treat the virus. Needless to say, alongside such collaboration and data sharing there was intense competition among countries and laboratories to be the first to obtain results.

The *public school* of open science is more concerned with sharing with the public than with experts, its goal being to make science accessible to a wide audience (Fecher & Friesike, 2014). This is important for researchers because, as pointed out above, research evaluation places increasing importance on 'broader impact' or 'societal relevance', that is, the impact of research on the economy, culture, society or policy. Fecher and Friesike (2014) distinguish two streams within the public school of science. The first stream is concerned with opening up the research process, allowing the public to play an active role. This stream is reflected, for instance, in the development of citizen science projects, where the public is usually enlisted to collect and produce data, thus participating in the research and contributing to knowledge production (see, for example, the citizen science platform Zooniverse); or in public participation through the crowdfunding of science projects of their choice (see Chapter 8).

The second stream of the public school concerns the need to make the products of research understandable to a wide audience. Puschmann (2014: 91) claims that scientists need to communicate their research clearly to the public to gain their support, and further adds that this support is necessary to get funding to conduct future research, to arouse public interest that may help to recruit junior researchers and to gain prestige. As Cribb and Hartomo (2010) argue, in order to face current challenges (e.g. climate change, poverty, ill health), sharing knowledge widely and making it accessible and easily understood by as many people as possible is essential. In the case of the Covid-19 health crisis, for instance, journals are making available not only relevant articles to be used by researchers, but also other digital genres for a broader audience, where the virus and its treatment are explained in non-technical terms (e.g. videos).

The literature about knowledge communication stresses that researchers should reach out to and engage with the general public by making their research comprehensible (Baron, 2010). Along these lines, Kelly and

Kittle Autry (2013) insist that making research available online does not necessarily mean that it is conceptually accessible to the general public. Given the importance of communicating research to non-specialists, researchers are increasingly called upon to use language that communicates information effectively to a wider audience. Several associations and foundations provide toolkits and resources to help researchers share their research with the public and enhance scientific understanding (e.g. the American Association for the Advancement of Science [AAAS], the National Science Foundation). Digital genres such as blogs or Twitter also help to make disciplinary knowledge understandable to the interested public, since scholars adapt their communication strategies to the information needs of the intended audience (see Chapter 7). As Fecher and Friesike (2014) remark, these digital genres are particularly useful for reaching out to the public using an informal style that differs from the style of conventional academic writing. 'Informal style' or 'informality' is a concept recurrent in the discussion of genres used to reach the general public and therefore needs to be defined here. Although 'informality' is a fuzzy concept, it tends to be associated with colloquial language used in everyday conversations and with the concept of engagement. As Hyland and Jiang (2017: 42) emphasize, however, informality is not 'just a colloquial style or the opposite of detachment' but 'the expression of a more personal tenor which implies a closer relationship to readers, a willingness to negotiate claims and a positive attitude towards subjectivity'. They add that informality also involves a set of distinctive features (e.g. 'I' and 'we' to refer to the author[s], 'you' to refer to the readers, direct questions) used to achieve specific rhetorical goals.

The *infrastructure school* is concerned with the technical infrastructure (i.e. software tools, applications and computing networks) that enables efficient research through the internet. Fecher and Friesike (2014) focus on two infrastructure trends: distributed computing, through which data-intensive research projects can be conducted by using the computing power of many users; and social and collaboration networks for scientists, which promote interaction and collaboration between physically dispersed researchers. It should be noted, however, that although these 'open' technologies certainly support open science, some scholars warn that they may also help to reproduce or intensify inequalities in knowledge production (Albornoz *et al.*, 2019); for instance, in some projects between Northern and Southern partners, equitable collaboration may be difficult because the Northern partner often controls the project resources and, therefore, may be in a better position to establish priorities.

Finally, the *measurement school* of open science is concerned with alternative measurements of impact. Given the crucial influence of the impact factor on a researcher's reputation and career, there is a need for alternative or complementary standards to measure this impact.

Advocates of this school suggest that the impact of forms of publication other than the journal article (e.g. blogs) should also be assessed and that impact assessment should also take into account the coverage of a scientific contribution on the web, for example, bookmarks, number of views, readings or downloads and Tweets (Fecher & Friesike, 2014). However, it may be the case that although digitalization may promote new measures of quality and impact, it has also reinforced particular evaluation regimes. As Nygaard and Bellanova (2018) argue, the online platform Google Scholar, a bibliometric measure that is freely available, makes it easier to monitor performance, and thus does not challenge but supports the use of citations to measure impact.

3.4 The Impact of Digital Media on Scholarly Communication

Digitalization potentially enables researchers to increase their participation in the scholarly arena and offers new channels for disintermediated interaction with the public in general. It has radically changed research activity, scholarly publication, assessment of research, knowledge dissemination and the relations between the actors involved in all stages of research. In this section, we outline the main impacts of digital media on knowledge production, communication and dissemination.

(1) Rapid, timely and low-cost dissemination of (published) research

The web reduces time frames and can enable rapid communication of results and ideas and immediate interactions among scholars (Bjerglund & Söderqvist, 2012; Rowley-Jolivet & Campagna, 2011). In addition to providing access to traditional genres such as journal articles and abstracts, the internet facilitates rapid access to other scholarly resources, such as working papers and pre-prints of journal publications, including those behind a paywall. In a similar fashion, web portals maintained by scholarly societies enable fast dissemination of research across disciplinary communities worldwide (Pérez-Llantada, 2016; Royal Society, 2011).

Many diverse audiences and stakeholders can access the generated knowledge more easily than in the past (Bjerglund & Söderqvist, 2012). Researchers in many contexts can access platforms where they can freely share their ideas, for instance through pre-prints, with no gatekeepers stopping or delaying their publication, which increases their visibility and their research impact, and helps to establish 'priority of discovery' for ideas and claims (Vale & Hyman, 2016). Although certain measures and pressures may limit the free access to content (e.g. paywalls, competition and publishers' exclusive rights), researchers today have more tools to disseminate their research more easily and rapidly than in the past. It is important to remember, however, that not all scholars have equal opportunities to disseminate their research or to access others' research.

Some researchers working in peripheral regions may lack the material resources (e.g. computer hardware or software, reliable internet connectivity) to take advantage of the possibilities for participating in global academia that the internet offers.

(2) Disintermediated communication with the public

Digital media can also make it easier for researchers to communicate directly not only with peers but also with the public (López-Pérez & Olvera-Lobo, 2016; Puschmann, 2015a; Trench, 2008), the public being understood here as all non-academic audiences, including the general public, governments, organizations, industry, business people, students and the media. Disseminating knowledge to the public is important because the results of research, especially in some disciplines, may have important societal implications (e.g. for education, poverty, industry and innovation, climate research or health-related research) and the public must be able to take informed decisions (Puschmann, 2014). However, until the advent of social media, it was difficult for researchers to share their work with the public without the mediation of other stakeholders (e.g. journalists, publishers, the media). The use of digital media to communicate directly with the public offers immediacy and enhanced visibility for academic research. Doubtless, the internet has made scholarly communication more publicly visible, which means that the work of researchers is getting more attention from the public than in the past. At the same time, it is becoming increasingly scrutinized and evaluated by the public (Puschmann, 2015a). Moreover, by facilitating communication with the public, digital genres help researchers to perform the role of public intellectuals, and discuss new research publicly in order to 'influence political, economic, social or cultural issues' (Kyvik, 2005: 290). Yet, we should bear in mind that dissemination of information without vetting may result in disinformation and in the public's difficulty in assessing whether findings are the result of evidence-based research and not simply belief-driven discourse (e.g. fake news on climate change, or vaccine conspiracy theories).

The internet has also blurred the boundaries between expert (specialist) audiences and non-expert publics (Trench, 2008), resulting in what Harmon (2019) refers to as 'a dual implied audience', and Luzón (2013) as a 'diversified audience'. The possibility of communicating simultaneously with diversified audiences has led to what has been conceptualized as 'context collapse', or the 'flattening out of multiple, distinct audiences (...) such that people from different contexts become part of a singular group of message recipients' (Vitak, 2012: 451). Context collapse poses important challenges in scholarly communication because, as boyd (2002) stresses, when it happens, internet users have to manage their interactions complying simultaneously with diverse sets of norms.

(3) New types of interaction

In addition to making research accessible to new publics, the web affords two-way communication and new types of interaction with both other researchers and the public (Bjerglund & Söderqvist, 2012; Puschmann, 2015a). Digital genres such as Twitter, for example, enable the audience to interact with researchers by contributing comments, asking questions, retweeting or liking tweets, among other possibilities.

Social media platforms are recruited to support the demand for a more balanced and dialogic relationship between experts and the public (Puschmann, 2015a). The use of social media for communicating with the public promotes a dialogue-based view of knowledge communication and facilitates a shift in the role of the public from being passive spectators to be educated to being participants who can contribute to and evaluate research (López-Pérez & Olvera-Lobo, 2016; Minol *et al.*, 2007; Puschmann, 2015a). Digital genres can provide useful tools for conducting different types of citizen science projects (see Bonney *et al.*, 2009). An example can be found in the online game Foldit, in which participants contribute data to scientific research by solving puzzles in order to predict protein structures.

Focusing on collaboration among researchers, Veletsianos and Kimmons (2012: 766) argue that, with the development of participatory online technologies, a new form of scholarship is emerging which they call 'networked participatory scholarship' and define as 'scholars' participation in online social networks to share, reflect upon, critique, improve, validate, and otherwise develop their scholarship'. However, there are other dynamics at play, such as the pressure on scholars to claim authorship of research for evaluation purposes or the need to obtain funds for research, that create barriers to this open sharing and collaboration.

(4) Emergence of new forums and new modes of scholarly communication

As noted, the affordances of the web have enabled the emergence of new publication outlets (e.g. e-only journals, such as those published by the Public Library of Science, which since their inception harness the affordances of the digital media, and tend to support open access) and new genres and platforms for informal communication with peers and with the public (tweets, blogs, wikis, academic social networking sites). Most of these genres are multimodal, extending the types of semiotic resources available to scholars to disseminate knowledge, and making it easier for researchers in some disciplines to meet the demands for reproducibility of research (e.g. through videos demonstrating procedures).

However, the uptake of these informal genres has to date been uneven among researchers, possibly because despite their benefits (e.g. public engagement, visibility, interdisciplinary networking and offering opportunities for communication), they also pose risks and involve challenges. Scholars' exposure to different audiences increases the chances of poor choices and missteps being noticed, which in turn may have negative effects on their academic reputation and career, and may lead to cybervigilantism and online abuse (Hildebrandt & Couros, 2016). Engagement with social media also results in an increased workload involved in creating texts for these platforms, which may cause conflict with the academic work-life balance (Huggett, 2019). Scholars' limited availability of time, and the need or wish to devote this time to activities that are recognized as 'academic work' may be the reason why some scholars refrain from writing informal genres.

(5) Increased integration and connection among the genres involved in the stages of scholarly communication

According to Lievrouw and Carley (1990), scientific activity is a cycle consisting of three possible stages: conceptualization of knowledge, documentation of research and results, and popularization of findings. At the conceptualization stage, communication is typically interpersonal and informal, with researchers sharing and discussing ideas in one-on-one or small-group interactions. At the documentation stage, communication is more formal, and is concerned with the production of a documented record of research results, such as journal articles or presentations at conferences. In the third stage of the cycle, popularization, research results may be communicated to society, although this does not always occur.

The archival, hypertextual and multimodal affordances of digital media enable the material connection of genres involved in these three stages, which results in the possible accessibility to complex genre assemblages, for example, online journal articles, technical reports, patents, media articles, blogs and microblogs (Pérez-Llantada, 2021b; Reid & Anson, 2019).

(6) Opportunities for network building, multidisciplinary research and geographically distributed research

Web-mediated collaboration and knowledge exchange via digital media have important benefits for researchers today. First and foremost, they support research networking within and across disciplinary communities. Through social media, researchers can multiply opportunities to engage in informal conversations with colleagues in other locations and working in other disciplines, which helps to strengthen and extend their networks (Gruzd *et al.*, 2012). Social media can be useful tools for the informal communication of 'invisible colleges' (Crane, 1969), that is, communities of researchers that have common research interests, work together and cite one another, but are not necessarily affiliated to the same research institution or co-located. Research in the 21st century is widely distributed geographically, with teams of researchers working in different institutions interacting and collaborating over long distances (Lillis & Curry, 2010; Luce & Di Giacomo, 2003) thanks to the affordances of new technologies. These technologies enable the use of shared infrastructures and the availability of prepublished data and research results in open access repositories.

We should add that social media facilitate the creation of virtual communities, which are not restricted to experts but may also include members of the interested public, thus offering researchers 'new ways to participate in wider global debate, with diverse audiences' (Scanlon, 2014: 14). Social media are therefore important tools not only for knowledge dissemination, but also for seeking connections, strengthening relationships and informal networking (Gruzd *et al.*, 2012; Puschmann, 2015a). An important concern here, and an issue that needs addressing in future research, is the extent to which this affordance is beneficial for researchers in 'peripheral' contexts, who may use social media to establish and maintain connections and to participate in informal communication with other members of the disciplinary community. Walsh and Roselle (1999) found that for researchers who moved to 'peripheral' institutions after being trained at 'center' institutions, the use of information and communication technologies made it easier to maintain their social ties, but this might not always be the case for other peripheral scholars and might not be the case at the present time with the increasing diversification of digital communication means and tools.

3.5 The Question of Language(s) in Online Knowledge Communication

As scholars from different linguacultural backgrounds are generating, sharing, communicating and accessing content online, and they may use different languages for these purposes, the issue of language use and multilingualism becomes germane to understand digital genres for knowledge communication.

It has been widely accepted that in peer-to-peer knowledge communication, many multilingual researchers' participation in global academia has been supported by the use of what many see as an instrumental lingua franca, namely English. Although English as the language for international research publication purposes has raised numerous concerns about English linguistic dominance (Skutnabb-Kangas & Phillipson, 2011), in an increasingly networked context English can be a facilitating resource for international cooperation, enabling researchers from diverse linguacultural backgrounds to create joint knowledge and then share and disseminate their results to the international expert community (see e.g. Corcoran *et al.*, 2019). Or, in Swales's (2019b) words, with English as an instrumental, mediating tool, researchers can strengthen their cooperation and new interdisciplinary fields emerge that draw on shared knowledge to advance scholarly inquiry. However, as already noted, it is not English alone that facilitates participation in collaborative knowledge production but also the availability of material resources, such as infrastructure, internet connectivity, bibliographical resources and appropriate wages (Canagarajah, 2002; Curry & Lillis, 2010; Salager-Meyer, 2014).

There is also prima facie evidence of multilingual communication in the academic world. If we delve into the everyday activity carried out in academic and research settings, we find that researchers draw on multilingual genre repertoires, for example, when they search for online scientific literature, prepare proposals for national and international funding organizations, travel for job-related purposes (e.g. fieldwork, conferences), exchange ideas and results with peers at the local or national levels, or prepare presentations for national and international conferences or informal speeches for local research outreach events (Ferguson *et al.*, 2011; Lillis & Curry, 2010; Muresan & Pérez-Llantada, 2014; Pérez-Llantada, 2018). Despite the importance of English in academia, its role as a language for sharing and disseminating academic knowledge online needs to be discussed in relation to other languages, given the increasing multilingualism on the internet and the diversity of audiences that can be reached online. As many internet users work with multiple languages (see also, Pérez-Llantada, 2021b), we find it useful to draw on Wendel's (2005: 51) ecological approach to language, because it acknowledges the complex relations between speakers' languages and the environment, including the online environment. This overt recognition of such relations brings to the fore the role of the internet in multilingual communication, and in knowledge sharing among linguistically and culturally diverse audiences.

As in other academic contexts, the use of English as a shared language in digital environments can facilitate multilingual scholars' participation in the international scholarly arena. In the context of open science, digital genres may draw on English as a medium of communication to construct, share and exchange new disciplinary knowledge. By way of illustration, the open lab notebook genre (Carter-Thomas & Rowley-Jolivet, 2017) and the open peer review report genre (Breeze, 2019) enable wider access to new knowledge through a shared lingua franca. Open digital genres relying on user-generated content (i.e. content that can be easily uploaded online by users), such as research blogs, offer more opportunities for multilingual scholars to disseminate their research globally, as these genres do not usually involve gatekeepers controlling, for example, the linguistic accuracy of the texts written in English (see Luzón, 2018b). English as a lingua franca (ELF) researchers argue for the need to accept linguistic practices that diverge from those deployed by Anglophone scholars (Mauranen, 2018; Mauranen *et al.*, 2020), which is increasingly the case in some online genres (see Luzón, 2018b, 2018c). Canagarajah (2018: 37) uses the term 'spatial repertoires' to refer to multilingual researchers' use of a given language in a particular situational space, to support 'meaning making and communicative success', even with non-standard varieties of language.

English is, however, only one of the languages used by multilingual scholars in their internet-mediated activity. They may engage in more than one language, for example, to share research data and publications in pre-print repositories, to maintain research collaboration via email on local, national, cross-border and international scales, to crowdfund projects, to post research updates on researchers' group websites or to blog followers with updates, news and events (Kjellberg, 2014; Luzón, 2017; Mahrt & Puschmann, 2014; Mauranen, 2013). In sum, these various activities involve multilingual practices that have received little attention in the scholarly literature.

One issue that also needs to be raised is that of researchers' language choice on the internet, with the affordances and constraints that particular language choices involve. Language choice is a complex decision for scholars from many linguacultural environments, with pragmatic reasons to use English and other languages (see Kuteeva & Mauranen, 2014). When composing digital genres for internal communication among peers at the international level (e.g. online articles or online laboratory notebooks), scholars may choose English because it is a prestige language, required in most high-impact journals, and a language that enables them to gain international visibility. However, researchers also engage in digital writing to make their research accessible to other stakeholders (e.g. local practitioners in the field who can apply the results of research, policymakers, students, lay public). Communicating with local audiences requires the use of the local language (Sajeev *et al.*, 2019). When composing some genres (e.g. research blogs and Twitter), scholars may decide to use only English, their local language, or use multilingual repertoires to communicate simultaneously locally and internationally (Han, 2019; Kjellberg, 2009; Luzón, 2017, 2018b, 2019c). For instance, English and other languages may be alternated in different ways in academic blogs and tweeting (Luzón, 2019a; see Chapter 7) and in researchers-followers interactions in crowdfunding projects (Pérez-Llantada, 2021b). This alternation of languages enables researchers to participate in different communities, to adjust their languages(s) to heterogeneous audience(s) and deploy their semiotic resources for different purposes. Digital genres and platforms enable scholars to use tactics to disseminate their research simultaneously to multiple audiences. For instance, research groups may use their websites to publish non-technical texts written in the local language to explain research reported in their English-medium publications, while at the same time providing a link to these publications (Luzón, 2019c), or they may produce online video documentaries about their research both in English and their local language (Sebastian & Giménez, 2016).

3.6 Concluding Remarks

In this chapter, we have provided an overview of some of the factors that shape researchers' academic activity in the 21st century and that influence knowledge communication practices in a number of ways. The central issues reviewed in this chapter are the influence of neoliberal economies on research and academic communication; the need for many researchers to publish quality research, and to achieve visibility and international impact; the increasing demand for openness in research and scholarship and its effects on the way knowledge is produced and communicated; the emergence of new genres spurred by the impact of open science and the possibilities offered by Web 2.0 technologies for scholarly communication; the need to explore the use of English for knowledge communication in relation to other languages on the internet; and researchers' choice of language(s) to reach diversified audiences.

We have argued that technological advances have enabled the emergence of new digital genres for knowledge dis nation and explained how these genres enable researchers to adapt their communication strategies to meet new socioeconomic demands and address changes in research culture. In the following chapters, we discuss what scholars do with digital genres, focusing first on how digital genres enable scholars to negotiate their identity and make their research visible (Chapter 4). However, even though digital genres have an important role in facilitating researchers' activity and making research processes and outcomes more accessible to various publics, in the book we also consider critically whether the potential of these genres to democratize research is being fully realized.

Note

(1) Science Europe is an association representing European organizations that fund and perform research.

4 Performing Multiple Identities and Enhancing Academic Visibility

4.1 Introduction

Identity construction is integral to knowledge production and communication, as scholars make particular discourse choices to negotiate their academic identities in the genres they produce. Genres are, as Bazerman (1994: 79) puts it, 'ways of being', and therefore allow scholars to enact specific identities. Thanks to the affordances of the medium (e.g. hyperlinking, multimodality, wide reach), digital genres enable scholars to use a wide variety of semiotic resources to construct their identities, to adopt new social identities and to enhance their visibility. Digital genres and online spaces have become quintessential for identity construction, self-promotion and reputation building for academics and researchers. Users can employ simultaneously several platforms for self-presentation and participate in various online communities, which enables the performance of multiple identities (Darvin, 2016; Kimmons & Veletsianos, 2014), some of which are not available to academics in more formal contexts. The combination of a variety of semiotic modes (e.g. images, sounds) afforded by the digital medium facilitates the construction of these complex identities. Online platforms also support multilingual communicative interactions and translanguaging practices that index particular identities (Androutsopoulos, 2015; Chen, 2013). In addition, social media enable the display of a network of connections, which has an important role in building scholarly reputation (Herman & Nicholas, 2019). Yet, despite these new possibilities, the construction of online identities also poses problems and challenges for scholars when they engage in knowledge communication and dissemination practices (Barton & McCulloch, 2018).

This chapter first explores identity construction in online environments, focusing particularly on academic identity. It then presents a review of the existing literature on how academic identity can be constructed through several digital genres. Next, it discusses the role of language choice in the negotiation of online identities by multilingual scholars. Finally, the last section presents a research case study which analyzes the blog as a genre exploited by research groups for identity performance. Although all online texts produced by scholars contribute to their visibility and thus to their identity footprint, the study in this chapter focuses on blogs because they are a genre particularly suited to self-promotion and community-building (Luzón, 2017; Miller & Shepherd, 2009).

4.2 The Discursive Construction of Academic Identity Online

4.2.1 Constructing online identity

Current research conceptualizes identity as socially and discursively constructed: identities emerge and are negotiated through interaction with others and participation in specific social practices and communities. Identity is recognized as complex, multilayered and dynamic (i.e. continually redefined to accommodate shifting goals and different contextual expectations/settings) (Flowerdew & Wang, 2015). This conceptualization is reflected in the framework proposed by Bucholtz and Hall (2005) for the analysis of identity construction in linguistic interactions (which synthesizes work on identity from several traditions, such as positioning theory). They define identity as 'the social positioning of the self and the other' (Bucholtz & Hall, 2005: 387) and conceive it as 'a relational and socio-cultural phenomenon that emerges and circulates in local discourse contexts of interaction' (Bucholtz & Hall, 2005: 586). As identity is constituted in social interactions, it includes not only macro-level demographic categories (e.g. gender, age, social class), but also positions created through interactional work in particular situated contexts. An example of a resource to index situated identities in online interactions is the use of specific languages or varieties, or code- and script-switching (Tagg & Seargeant, 2012). Bucholtz and Hall also emphasize the relational nature of identity. Identity construction is not an individual process but an intersubjective one, involving several complementary relations such as similarity and differentiation. Finally, they highlight that any given construction of identity is partial, since it changes as interaction unfolds and in different discourse contexts. Bucholtz and Hall's (2005) concept of identity as *emergent through social interaction*, *relational* and *partial*, provides an appropriate framework to understand how identities (including academic identities) are constructed and negotiated in digital spaces, as the review of the literature in this chapter discusses.

Focusing on research on how identities are constructed online, two influential theories – which accord with Bucholtz and Hall's (2005) framework – are Davies and Harré's (1990) positioning theory and Goffman's (1959) idea of identity as performance and selective self-presentation. Goffman (1959) considers that in a given context of interaction, people play out different parts of the self for specific audiences, trying to project a credible image: through performance, individuals endeavor to construct identities that are consistent with audience(s) expectations. When individuals interact with others, they seek to create desired impressions and highlight positive ideas of themselves to their audience. Researchers have emphasized the social and multifaceted nature of online identity (Barton & Lee, 2013; Davies & Merchant, 2007): participants in online interactions are networked individuals, members of different communities connected to each other, performing for different audiences and in varying contexts. Therefore, they may present different identities in accordance with their assumptions about the norms and about the audience's reactions in each context. As stipulated by positioning theory, individuals select particular registers, words and non-verbal cues to create subject positions and signal themselves as members of particular groups (Davies & Harré, 1990). Online identities are therefore the product of intentional self-representation, even if some discourse choices that individuals make are not conscious, but a result of internalized practices of interaction.

The affordances of digital environments (the possibility of reaching diverse audiences, multimodality, networking) shape the nature of online identities and the mechanisms through which they are constructed. Research addressing identity in digital media has explored the self-presentation strategies performed in these media and the resources available in various digital settings (Androutsopoulos, 2015; Chen, 2013; Dennen, 2009). These studies show that writers use a range of semiotic resources (e.g. visual, audio, interactive resources) to position themselves and create their online identity, reflecting that communication is a multimodal, and not only a linguistic, process. The construction of online identities has been characterized as a process of careful and deliberate selection and assembly of semiotic resources (Darvin, 2016). Rudolf von Rohr *et al.* (2019), for example, examined the positioning strategies employed by health professionals in digital spaces and showed the important role of visuals in the creation of credible expert identities. In addition to the wide range of available semiotic resources in online media and the possibility of reaching wider and more diversified audiences, another affordance of online environments, particularly relevant for academics, is networking, which facilitates the development of social and relational identities. Participants in web-based networked environments can display networks of connections (Donath & boyd, 2004), which contribute to the construction of identity through identification with and social affiliation to particular groups or communities. Marwick and boyd (2011: 129) use the term 'networked audience' to refer to the imagined audience consisting of people who are 'connected not only to the user, but to each other, creating an active, communicative network'. These audiences, consisting of real or potential consumers of digital content, are 'imagined' by the individual in order to make choices (e.g. language, style) for their online identity presentation.

4.2.2 The construction of online identity and reputation in academia

In the academic context, there has been a shift in the ways in which scholars present themselves and their work publicly: they are harnessing the range of affordances of online genres (e.g. web pages of their institutions, social networking sites, academic blogs, microblogs) to create online personas, develop professional identities and claim a place within their disciplinary community and outside academia.

The concept of 'community' is particularly relevant to understanding the development of academic identities through genres. Hyland (2011, 2015a) emphasizes that scholars construct their identities through social interaction in their communities. He claims that when adopting community-specific discourses, values and rhetorical practices, academics identify themselves as members of a particular community and position themselves in relation to this community. In most written genres, academic identity is related to the construction of a 'credible' academic persona by using the discourses and language forms of the community (Hyland, 2011). The genres of the discipline enable researchers to construct themselves as competent members of the community and at the same time express their individuality (Hyland, 2015a).

The building of an academic identity online involves intentional practices to align with the values of the community. Research has emphasized that scholars exercise agency through deliberate choices of symbolic resources to perform their academic identity online (Hyland, 2011, 2012; Lupton *et al.*, 2017). Marshall *et al.* (2017) use the term 'persona' to refer to a strategically developed public identity. This persona is 'a complex and varied combination of professional, private, personal, teacherly, networked public identities that individual academics attempt to manage' (Marshall *et al.*, 2017). The intentional self-representation of academics online involves making several decisions (Lupton *et al.*, 2017), such as what platforms to use and how to use them, what biographical information to post on digital spaces, what images to choose as portraits, what content to contribute, what relations with other users to make public (Lupton *et al.*, 2017), what languages and discursive features to use. In line with this point, Hyland (2012) describes the composition of academic homepages as a process involving the deliberate selection and arrangement of symbolic resources to construct an identity in accordance with the expectations of the community.

The construction of a digital identity is an important activity in today's academic world as a strategy for reputation building (Barbour & Marshall, 2012; Herman & Nicholas, 2019). Herman's (2018: online) definition of scholarly reputation emphasizes the importance of *visibility*, which can be facilitated by digital genres: 'Scholarly reputation is the expert appraisal of a *scholar's standing in their collegial reference group*, which is collectively determined on the basis of their research achievements', and is built by making one's research activities (i.e. activities involved in producing, disseminating and evaluating knowledge) and academic achievements visible so that they attract the attention of those who can judge their value. Herman and Nicholas (2019) describe several digital practices that contribute to building scholarly reputation. Some involve disseminating research results formally and informally to various audiences (the disciplinary community, the wider scholarly community and the public), such as disseminating research results formally in online journals; disseminating manuscripts and pre- or post-prints via repositories, personal websites or academic social networking sites (ASNSs); blogging; live tweeting from conferences; or posting recorded lectures or videos on social media. These practices help scholars to achieve visibility and scholarly impact within and beyond their discipline, to obtain peer recognition and to enhance their digital identity. Using digital genres, such as Twitter, to share others' work is another practice with a reputational purpose (Stewart, 2015a, 2015b). This practice increases scholars' visibility to the authors whose content they have shared – by using the @ mention feature of Twitter to explicitly address the post at these authors – and displays their own interest and expertise to their networks, creating ties with other scholars with shared interests. Scholars who circulate content within these networks become more visible to each other.

Studies on the identity performed by academics in online platforms have paid attention to its fluid, complex and fragmented nature. Jordan (2017) conceives online academic identity as a spectrum between the personal and the professional, with different sites supporting different academic identity fragments, that is, occupying different segments of the spectrum. Barbour and Marshall (2012) found five styles of online academic personas that, to a certain extent, relate to different online platforms and genres: the formal self, the networked self, the comprehensive self, the teaching self and the uncontainable self. The formal self, characterized by the static presentation of achievements and expertise and a lack of interactivity, is commonly seen on faculty institutional web pages and professional networking sites. The networked self still focuses on the academic's professional life but it is interactive and can be seen on platforms that facilitate two-way communication and network formation. The comprehensive self, which smoothly mixes professional and private aspects, operates on platforms such as blogs or Twitter. The teaching self, or teaching persona, is seen on platforms that have features that facilitate teacher-student interaction intended to enhance teaching and learning (e.g. Facebook groups or pages associated with courses). Finally, the 'uncontainable self' is not controlled by the academic themselves, but by another person (e.g. a fellow researcher, a student) who can frame the academic's persona positively or negatively (e.g. by means of online criticism of their work).

The notion of the 'uncontainable self' brings to the fore the fact that although digital genres offer many affordances for the construction of new and more complex identities and contribute to increasing academics' visibility, these same possibilities pose challenges and raise concerns for researchers, especially for those who are not comfortable with self-exposure or self-marketing. Barton and McCulloch's (2018) study of the writing practices of academics at three English universities reveals some of their concerns regarding the creation of online identities, that is, the feeling that they are expected (even forced) to maintain an online identity, both on their university's website and on other sites outside the institution, the ease with which their lives are searchable, or the tension between the values of academia, where the focus is on the quality of the work, and the promotional purpose of these genres. Other scholars have also drawn attention to the difficulty of self-presentation on social networking sites where personal and professional audiences are merged (Veletsianos & Kimmons, 2013), and to the negative consequences that missteps or poor choices online may have for the reputation of scholars (Hildebrandt & Couros, 2016).

4.2.3 Digital genres to construct academic identity

This section presents a review of the existing literature on the construction of academic identity online, focusing on how various online genres and platforms are used for this purpose. This is not a comprehensive review, since there is an extensive and fast-growing body of research on online academic identity. The studies discussed here serve rather to illustrate how the affordances and constraints of different digital genres determine the identities that researchers can construct.

Homepages and websites were one of the first digital genres used for identity construction in academia. Research institutes, faculties, research groups and individual scholars have been using home pages and websites for promotional purposes and identity work. Hyland (2012: 311) characterizes the personal homepages of academics as a medium to construct credibility as experts and declare their academic standing. Drawing on Davies and Harré's (1990) positioning theory, he examines the ways in which visual, textual and hyperlink features are used to construct identity in academic web pages and concludes that these semiotic resources interact to manage a 'rhetorical machined identity likely to appear credible to others' (Hyland, 2011: 296). Hypertext links contribute to the creation of a virtual community and indicate allegiances, thus helping academics to position themselves within the disciplinary community. These pages display mostly professional biographies and information on research interests and publications, which helps scholars to build their reputation and construct their identity as experts. Similar studies such as Dumont and Frindte's (2005) research on psychologists' homepages, reveal the prevalence of information about research activities. If we consider academic identity online as part of a spectrum from the personal to the professional, with different sites occupying different segments of the spectrum (Jordan, 2017), homepages tend to support the performance of professional identities, enabled mainly by the display of research output.

Homepages and websites are also being leveraged by research groups for reputation building and visibility (Lorés-Sanz, 2020). Our study of the websites created by a set of research groups affiliated to a Spanish university (Luzón, 2019c) revealed these websites as promotional genres through which the groups publicize themselves in front of various audiences (i.e. peers, research funding bodies, practitioners, students, general public). These sites are used to construct the group's academic identity and build its reputation through public visibility of its research output, activity and networks. The identity of the group is constructed not only by providing and embedding different types of information about the group, but also by linking to and aggregating the information about the group and its members on various online platforms, including their publications, video channels, profile pages in social academic network sites and accounts on social media.

The blog is another digital genre that has been robustly studied in relation to identity construction. Blogs can be spaces for self-expression, self-promotion, group relationship and the construction of social identities (Dennen, 2009; Luzón, 2018a), where academics can 'author the self' and forge their identities through connections with others (Davies & Merchant, 2007: 192). The features of blogs and their technological affordances make them a fertile space for the development of academic identities. Blogging academics construct their identity by writing a chronologically ordered series of posts, which enables an ongoing and iterative process of subject positioning. By providing cumulative, personally meaningful and diverse information about various aspects of the blogger's academic (and sometimes personal) life, blogs facilitate the performance of multifaceted and dynamic identities (Graupner, 2010). Blogs also support the smooth combination of various modes and enable identity performance and self-presentation through a variety of media (Kirkup, 2010: 76; Merchant, 2006). Finally, the participatory nature of blogs supports the creation of a social identity (Boyd, 2003; Merchant, 2006) and the desire of individuals to affiliate, and thus facilitates the creation of new social groupings. In this sense, Merchant (2006: 241) points out that blogging supports the emergence of networks which are both global and local, by facilitating interaction between local and geographically dispersed individuals who share interests or are members of a community of practice. By harnessing these affordances of blogs, researchers can increase their visibility (by showcasing their research activities and output) and enhance their reputation as knowledgeable members of a disciplinary community as well as collaborate and interact with others.

In line with the claim that the construction of academic identities serves to position researchers within a community (Hyland, 2011; Lupton *et al.*, 2017), several studies have shown that the way scholars use various semiotic resources and strategies to construct their identity in blogs is influenced by the norms and values of the community to which they belong, or desire to belong. Dennen (2009) identifies the elements used by a community of academic bloggers to develop and maintain their academic identities, such as including name and blog title, profiles, post contents (i.e. sharing details about their lives and interests), voice and linguistic choices, affiliation (expressed through blogroll[1] links, badges and links to other websites) and visual design (e.g. photos of themselves). Dennen supports a developmental view of online academic identity as being influenced by community norms. She points out that blogging identities are constructed by combining verbal, visual and structural elements and that the way these elements are used is affected by the norms of the community to which the blogger belongs. The function of disciplinary values and professional norms to affect identity construction in blogs was identified by Kedrowicz and Sullivan (2012). They used qualitative content analysis to investigate how engineers presented themselves in 14 top engineering blogs. The fact that the engineering culture tends to emphasize objectivity, reason and distance and depreciate subjectivity and engagement is reflected in the strategies used by engineering bloggers to establish credibility. Mewburn and Thomson (2018) show that the potential of blogs for indexing membership to a community is particularly important for novel researchers. Focusing on the practices of doctoral students, they describe blogging as a technology of self-formation, a site for the construction of community and disciplinary belonging. Blogging helped the doctoral students in their study formulate their ideas, learn to communicate as researchers in an academic context and develop an understanding of what it means to be a researcher.

In addition to websites and blogs, social media are also being increasingly used to engage in identity performance. Twitter, for example, is used by academics for several purposes significant in construing academic identities, such as presenting and publicizing professional output, informing readers about their social activities or making their relationships with others visible (Budge *et al.*, 2016; Shah, 2015; Veletsianos, 2013). Research on academics' use of social media shows that identity is constructed by combining professional and non-professional/personal aspects. In her study of the use of Twitter by academics, based on semi-structured interviews with 28 academics at the University of Sheffield, Shah (2015: 297) found that participants constructed their identities by mixing professional and personal elements, projecting both their academic reputation and their sociability, to create academic profiles accepted by the community. They achieved this aim through several tactics (e.g. sharing interests and academic work, providing images, offering suggestions to others' request for assistance or displaying their character by sharing content that shows them as being serious about their work, friendly, entertaining, approachable or interesting, among others) intended to present their 'true-self' by displaying fragments of their identity that would be accepted by others.

Research on social media has also highlighted the importance of networked audiences in the definition of online academic identities (Budge *et al.*, 2016; Jordan, 2017). Weller (2011: 4) notes that in this digital, networked world, researchers' networks and online identity may be more important in defining them than their institution. Stewart (2015a) used ethnographic methods to analyze the practices of scholars on Twitter and found that, unlike academic homepages, social media can foster connections and collaboration between individuals and the development of academic identities and networks as individuals, rather than only as members of particular institutions. This research also emphasized the partial, fragmentary and intentional nature of social media identity: this identity is a constellation of incomplete, transitional and intentionally developed 'identity fragments', and academics may perform different personas on different sites (Jordan, 2017; Kimmons & Veletsianos, 2014). However, although academics tend to use several platforms to display their digital identity, most research has focused on identity construction within a single genre or platform. Thus, Bukvova (2011) argues for a more holistic approach to the study of academic online identity, one that considers the online presence of academics across several platforms. She accordingly analyzed scientists' profiles on institutional and private web pages, social networking services, blogs and microblogs, paying attention not only to 'profile instances' (i.e. 'the content provided as a part of a single profile on a particular platform') but also to 'profile networks' (i.e. 'network of profile instances belonging to one scientist' which 'can be connected with hyperlinks') (Bukvova, 2011, 'Results', para. 2). The hyperlinking affordances of digital genres make it possible, therefore, to connect the fragments of an academic's online identity.

Given our focus, we close this review by noting an aspect of research on academic identity construction online that has been overlooked: the use of language choice for this purpose by researchers from non-Anglophone backgrounds. Most studies reported above have analyzed English-medium texts written by researchers in English-speaking environments. A few studies have explored the practices of researchers from non-Anglophone backgrounds but rarely making reference to the language of the texts. Bukvova (2011), for instance, analyzed 79 online profiles of 15 German scientists, but did not provide information on the language used by these scientists to provide content. The following section focuses on the role of language choice and discusses the few studies that, to our knowledge, have analyzed its relation to scholars' identity work in digital environments.

4.3 Multilingualism, Language Choice and Identity in Online Environments

The affordances of online environments support specific multilingual practices, which can be exploited by users for identity performance and negotiation. Research on the role of multilingualism on the internet has shown that languages are strategically chosen to negotiate and index identity and to develop a sense of belonging (Androutsopoulos, 2007; Han, 2019). The choice of language seems to be related to the participants' intention to project a local or global identity and signal their affiliation with a specific community (Barton & Lee, 2013). In online environments, English is a resource shared by many internet users that helps them to participate in various communities, but it interacts with other languages in various ways: multilingual users may mix English with their local languages/other languages, or alternate between English and their local languages or other languages (Leppänen & Peuronen, 2012: 388). Thus, translanguaging, or the use of all one's linguistic resources to construct meaning and perform identities, has become a salient practice to express a complex, hybrid identity (Androutsopoulos, 2015; Han, 2019). Translanguaging is exploited by language users to negotiate their 'glocal' (i.e. both global and local) identities and project new forms of identities that may go beyond those related to ethnicity, nationality or gender/sexuality (Lee & Barton, 2011). The concept of 'networked audiences' helps to understand how multilingual scholars use language online to construct their identities. The networked audiences imagined by users influence their linguistic choices, as they may select or combine specific languages to respond to the expectations of these audiences, or to target specific audiences in translocal spaces where contexts collapse (Androutsopoulos, 2014).

Research on the literacy practices of multilingual writers on social networking sites reveals that these users employ their linguistic repertoires and exploit the available semiotic resources to index multiple identities (Han, 2019). In her analysis of the use of Facebook by two multilingual writers, Chen (2013) found that they aligned themselves with particular identities by reappropriating the symbolic resources available on this site. Her study shows that social media can provide a 'hybrid third space' where multilingual writers can 'navigate across multiple languages, cultures, and identities, including their social, cultural, and professional identities through various literacy practices, language choices, and contents' (Chen, 2013: 143). Both participants in the study were international graduate students in the United States whose local language was Mandarin Chinese. Interestingly, they sought to project different identities and so they employed different strategies and resources to do so. One of the writers, Cindy, used linguistic resources to build relationships with different audiences: she used Mandarin to create bonds with other Chinese-speaking graduate students in the United States and index identities as a struggling student and as a teacher, and English to connect with her American friends and project herself as a legitimate member of the university. Jane (the other writer), by contrast, used mainly English – an additional language for her – to communicate online, and thus construct her identity as an international English user. Rather than linguistic resources, she used the content of her postings and other literacy and multimodal practices available on Facebook, e.g. status updates, hyperlinks, images, video sharing, to move between her multiple identities (i.e. as a citizen of China, and as a multilingual and multicultural cosmopolitan global citizen).

Of the few studies on the translanguaging practices of multilingual scholars in constructing their online identity, Han (2019) analyzed the linguistic and semiotic resources used by Chinese visiting scholars in their postings on WeChat Moment (a feature of the WeChat app where users can post texts, photos and comments, and share videos and articles). He found that using and mixing English and Chinese on this app is a strategy used to construct a bilingual identity. By translanguaging, Chinese scholars interacted with various networked audiences (potential addressees in China, a superintendent from the Chinese government, addressees in the host country) and presented themselves as both English learners making progress with the language and as global citizens. Other studies focusing on multilinguals' choice of language in blogs (Luzón, 2018b, 2019a) have demonstrated language choice as a strategy for audience selection and identity construction. Multilingual academic bloggers use language choice to transmit who they are and how they want their audience to perceive them. The mixing of languages (English and their local languages) in their blogs contributes to signaling complex and multifaceted identities: as multilingual scholars, as members of the international academic community and as scholars involved in local issues (e.g. politics).

In the following section, we discuss the results of a study of how research groups use their blogs to construct various identities (Luzón, 2018a).

4.4 Research Case Study: Identity Construction in Research Group Blogs Written by Multilingual Scholars

In many disciplines, research groups are the primary organizational unit in the practices of knowledge creation and they often use blogs as tools for self-presentation, self-promotion and identity management. In Luzón (2018a), we explored the identities performed by 12 Spanish research groups who write their blogs in English (or in English and the local languages) and analyzed the semiotic resources drawn upon in these blogs for identity work. When developing their research and communication agendas, these research groups had to take into account the Spanish

Strategy for Science and Technology and Innovation (2012), which established the general objectives and priority lines of action to develop R&D activities in Spain during 2013-2020. This strategy highlights the importance of the quality of research and its social, economic and international impact, and establishes the promotion of collaboration, the support for the internationalization of research and the dissemination of scientific culture in Spanish society as priority lines.

The corpus for this study consisted of 347 posts from 12 blogs written by research groups in different disciplinary fields (science and technology, medicine, social sciences and humanities). They are blogs authored by multiple authors, with posts by different members of the groups, mainly used to showcase their academic output and research activity, and to disseminate information related to this activity.

The findings show that research group blogs are reputational platforms that the groups employ to perform intentional acts of positioning, by posting carefully selected information about themselves and their academic lives in various modes. The groups in the study draw upon various semiotic resources to develop dynamic multiple identities and position themselves as (i) competent and active researchers; (ii) networked researchers within their disciplinary community; (iii) members of the international research community, involved in collaborative research, and of the local community; and (iv) civic scientists concerned with societal issues. Not all groups, however, perform all these identities in their blogs, and some identities seem to be much more important for some groups than for others. In what follows, we discuss how these identities are constructed.

Competent and active researchers. All the blogs in the corpus showcase the group members' research activities in order to project scholarly credibility and authority. Texts, visuals and hyperlinks are combined to provide information that works to create an expert and knowledgeable identity. For instance, they use: (i) texts to announce members' new publications and achievements and describe the content of the group's projects/research, upcoming events organized by the group, news of interest to the disciplinary community; (ii) pictures of group members at academic events, figures and graphics from members' research output, images of members' publications, and conference posters; and (iii) links to their own productions (e.g. online papers, abstracts in databases, preprints, posts, online presentations) to provide information on the group on other sites or platforms (e.g. member's homepage, group's homepage and press releases). A high number of posts in the corpus (21%) was intended to announce new publications and make them available, combining resources to increase their visibility. Example 4.1 presents a fragment of a post in a blog published by the Distributed System Architecture (DSA) research group. In addition to the links to the journal, to the institute with which they are collaborating, and to the article ('here'), the post includes an image of the first page of the article, providing visual evidence of the publication, and a summary of the publication, where the main results and their significance are emphasized in bold.

> **Example 4.1.** The International Journal of Parallel Programming has just made available online our latest work on approximate request processing in cloud online services. This is the result of our collaboration with the Institute of Computing Technology from the Chinese Academy of Sciences and it can be accessed here (http://blog.dsa-research.org/)

The DSA group describes itself on the 'about' page of its blog as a group which 'conducts *outstanding research* in *collaboration with international research centres and companies*, aiming to become a *research group for excellence* in advanced distributed computing' (our emphasis) and lists the content of the posts in the blog: 'Research activities and articles of the dsa-research group; description of new research projects, opinion, observations and discussion on scientific and technological challenges; announcements of calls for papers or conferences with our participation; participation in research events'. From this description, the blog seems to be for them a tool to project their identity not only as competent outstanding researchers in their field, which they do by showcasing their research output and activity, but also as members of the international research community and networked researchers, engaged in international collaboration.

Networked researchers within their disciplinary community. Although all the posts in the corpus signal in different ways how the groups belong to the disciplinary community, in some posts bloggers use more explicit strategies to express affiliation within a discipline and construct social bonds within the community. Research groups use a range of resources to display their collaboration networks, by providing, for example, information on collaboration or visits from other scholars, and pictures providing evidence of such collaboration. At times, group members also use these blogs to share personal experiences related to their work or provide a narrative to talk about academic events such as conferences or visits to other research institutions in a personal way, which contributes to creating a sense of community. Photographs are an effective semiotic resource used in these blogs to construct a social identity and position the group as part of the disciplinary community. Groups use different types of photographs to record and share common experiences, such as conference photographs, or more personal and informal photographs, such as those sharing non-academic activities with other members of the community, which reflect conviviality and relations of comradery or friendship. For instance, the post 'Pierre Çarçabal and Rolando Lozada visited our laboratory', on the blog by the Spectroscopy Group, includes the text in Example 4.2 and a photograph of the two visiting researchers and a member of the group, posing and smiling at the camera while having dinner at a restaurant. The blog of this group includes many posts informing of visits by other scholars, thus reinforcing its own description on the 'about' page of the blog as a group which 'has an extensive network of national and international collaborations'.

> **Example 4.2.** Pierre Çarçabal and Rolando Lozada from CNRS (France) visited our laboratory after several attemps. Its flight was cancelled by a strike. They spent three days in Bilbao where we were able to discuss future projects and programming the next visits.
>
> In the picture: Pierre Çarçabal (right), Rolando Lozana (center) and Emilio J. Cocinero enjoy the Basque cuisine (http://grupodeespectroscopia.es/MW/news/page/6/)

The blurring of the groups' public and private academic identities in some blogs (see Example 4.2) contributes to presenting the group as maintaining close social relations, which facilitate research collaboration, with other members of the disciplinary community.

Members of the international research community and of the local community. Groups also exploit blog affordances to construct their international identity by presenting themselves as engaged in international collaboration (see Examples 4.1 and 4.2) and using English to write their posts (in 97% of the posts analyzed). Among other purposes here, posts inform and provide graphical evidence about international projects or networks, publications co-authored with researchers from other countries, participation in international conferences, and PhD committees involving international members. A powerful tool to construct an international identity is the use of English. It helps multilingual scholars to construct their identity as insiders in the global research community and to gain visibility and recognition in the international academic community. In six blogs in the study, however, English coexists with local languages (Spanish or Catalan), which projects a hybrid glocal identity: posts in English signal the international nature of the group's research while posts in Spanish or Catalan, which deal with matters of particular interest for a local audience, project the group's local identity as researchers concerned with local issues, thus attracting local support and funding. For instance, the GeNeura group writes most of the posts in their blog (https://geneura.wordpress.com/) in English, but uses Spanish to write about their dissemination activities, for instance in the post 'GeNeura en la Noche de los Investigadores' (where they write about their participation in the European Researchers' Night). Another group that uses the local language in its blog is the Researching Gender in the Network Society (GENTIC) group. This group, which conducts research on how gender relations affect the development and use of technological innovations, describes itself on the blog 'about' page as being 'active in projects at regional, national and international level'. Although most of the posts are written in English, as they are used to inform about their research activity, Spanish is sometimes used for posts about issues of particular interest to the local audience. For instance, the post 'Las diferencias salariales no desaparecen. Aspectos a tener en cuenta en la celebración del Día Internacional de la Mujer' ('*The gender wage gap does not disappear. Aspects to take into account on the International Women's Day*'), which provides information on the gender wage gap in Spain, evinces their interest in local issues.

Civic scientists. Some groups also use their blogs to position themselves as 'civic scientists', that is, scientists who engage in dialogue with the public on important issues. To project this identity, they include posts about social issues (e.g. gender, environment, animal welfare) or posts reporting on their dissemination activities.

For instance, the GENTIC group, whose research on gender, information and communications technology (ICT) and science, technology, engineering and mathematics (STEM) has important societal implications, uses its blog not only to disseminate its research activities and output, but also to raise awareness about gender issues and project its civic identity as a group committed to, as stated in its blog, '[providing] new insights into emerging societal challenges by incorporating a gender dimension'. In the post 'PREZIosa initiative' (https://gender-ict.net/2015/05/preziosa-initiative/), for example, readers are encouraged to take action and participate in one of the group's initiatives that aims to raise awareness about breast cancer and collect funds to study, prevent and fight against the disease (see Figure 4.1). This post is written in English, probably to make it accessible to an international audience and to make the group's civic identity prominent also for that audience. However, the website for the PREZIosa initiative (which can be accessed through the link PREZIosa initiative), where the group provides detailed information on the initiative and on how readers can contribute, is in Catalan, clearly intended for a local audience (see Example 4.3, taken from the 'about' page of the website).

> **Example 4.3.** Els membres del grup de recerca GENTIC volem afermar el nostre compromís envers la millora de la qualitat de vida de les dones. Per aquest motiu, impulsem una acció que vol ajudar a:
>
> 1. Sensibilitzar a la població sobre el cáncer de pit.
> 2. Recollir fons perquè siguin destinats a diferents activitats per prevenir, lluitar i continuar amb la recerca sobre aquesta malaltia. (https://preziosagentic.wordpress.com/)

Other groups also project their civic identity by making readers aware of societal issues not directly related to their research. For instance the blog by the Global Energy and Environmental Economics Analysis

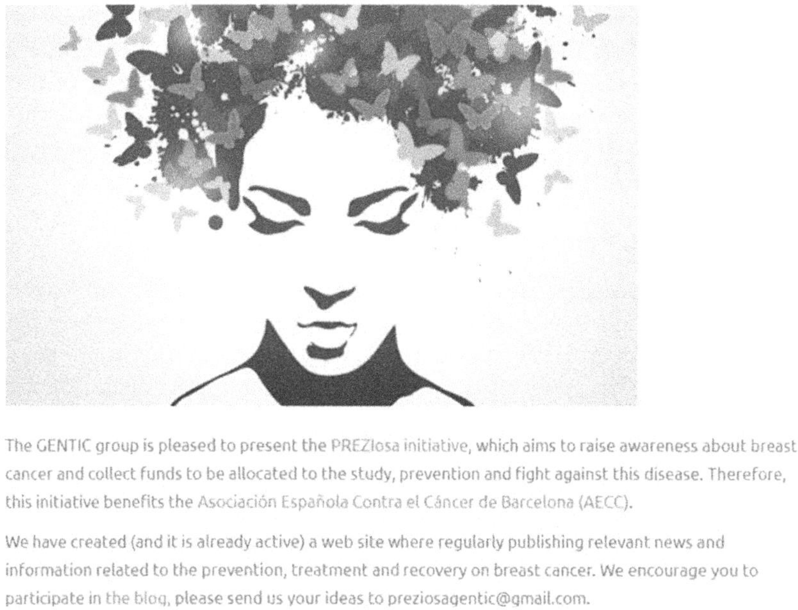

Figure 4.1 Screenshot of the post 'PREZIosa initiative' (https://gender-ict.net/2015/05/preziosa-initiative/)

Research Group (GEAR) includes the post 'GEAR presente en la V Gala Benefica de AUDA' ('*GEAR present at the 5th Charity Gala of AUDA*'), where they describe the objectives of this non-governmental organization (NGO) concerned with animal welfare and link to the NGO's website, which helps to signal the group's commitment to this issue.

In sum, research group blogs seem to be genres particularly suited to the construction of a complex multifaceted group identity by providing selective, cumulative and diverse information about the group as a whole and its individual members. This study supports Bucholtz and Hall's (2005) concept of identity as emergent through social interaction, relational and partial. The group identities constructed in the blogs are dynamic and partial, changing as new posts are added and new information is aggregated, and with every new post focusing on one or several aspects of the groups' identity. When composing the posts, the groups perform for different audiences, selecting information consistent with these audiences' expectations (see Goffman, 1959): international and local communities, local and national governments or the public. In agreement with positioning theory (Davies & Harré, 1990), groups select particular semiotic resources to position themselves as members of particular groups. All the groups in the study seek to position themselves as competent researchers, networked within their disciplinary community and involved in collaborative research at the international level, English playing an important role in projecting these identities. The local languages are used by those groups who also want to project identities as members of the local community and as civic scientists. The use of their blogs to perform these identities helps these groups to show that their research activity is aligned with the objectives and priority lines established by research funding agencies and research policymakers.

4.5 Concluding Remarks

In this chapter, we have explored how digital genres can play an important role in enhancing researchers' visibility and helping them perform multiple identities (e.g. as competent members of the discipline, as civic scientists). We reviewed the construction of online academic identity as a tool for reputation management in an increasingly complex academic context where reputation is built in multiple ways: through achieving visibility in the discipline, networking and disseminating research to the public. We have discussed how academic identity is constructed in online genres such as blogs through the selection of identity fragments that will meet various audiences' expectations. Finally, we have sought to bring to the fore how the digital medium affords opportunities for multilingual scholars to draw on a variety of semiotic resources and on their linguistic repertoires to construct multifaceted identities.

More research is needed, however, to better understand the role languages play in identity construction, specifically, what languages multilingual scholars choose (and why) when enacting these different identities. As online identities are fragmented, and scholars may use a range of digital genres (e.g. blogs, microblogs, websites) to reach and engage diverse audiences, the relationship between languages, the variety of digital genres used by scholars and their identities need to be studied in more detail.

Finally, it is worth pointing out that the possibilities for identity construction afforded by various digital genres can be a double-edged sword for researchers. At a time when academic and research activity has become increasingly reliant on digital media, scholars may feel pressured to project their identity in digital spaces, since many institutions expect them to be visible and promote their work online. There are many reasons, however, why some scholars would rather not engage in promoting themselves online, including a lack of time, interest and digital skills, a reluctance to market their work, the difficulty of managing an academic identity that is visible to various publics and the lack of control of public audiences and their reaction to scholars' online participation. Given the importance of building scholarly reputation in academia, public identities are crucial for academics and they need to be carefully constructed. The complex combination of professional and personal identities afforded by online participation may increase scholars' reputations (if identity is skillfully managed), but poor choices in self-representation may also have a negative impact on their reputations and lives. In addition, the lack of access to technology by some scholars means that they cannot take advantage of digital genres for greater visibility, recognition, credibility and trust-building purposes, limiting their digital identity to what Barbour and Marshall (2012) refer to as the 'uncontainable self', an identity created by others, which is beyond their control. In sum, while digital genres such as blogs can help researchers construct their reputation, they may also work against those researchers who are more reluctant to market themselves online or who lack the skills or material resources to compose these genres effectively.

Note

(1) List of blogs, usually placed in the sidebar, that the blogger wants to share.

5 Sharing Research in Progress with Peers: Online Laboratory Notebooks

5.1 Introduction

Web 2.0 technologies are changing research routines and the ways in which scientists interact and share results. Many digital genres have emerged or evolved from pre-digital counterparts in order to meet researchers' need or wish to collaborate with peers, communicate with them formally and informally and share data, protocols and results (e.g. blogs, microblogs, enhanced publications, open lab notebooks) (McGrath, 2016; Puschmann, 2015a; Wickman, 2016). These genres, which facilitate different types of collaboration (e.g. interdisciplinary, international), have contributed to changes in the nature of scientific research. Many kinds of research have become more collaborative, interdisciplinary and geographically distributed, with researchers increasingly working in (cross-national) teams, and with research groups collaborating with each other and sharing resources and new discoveries/data to solve complex scientific problems (Delfanti, 2008; Luce & Di Giacomo, 2003) (see Chapter 3).

In this chapter, we explore a somewhat underused genre intended to facilitate collaboration during the research process, namely, the open laboratory notebook (OLN). This genre has emerged to respond to the demands of open science for sharing the whole research process. OLNs allow researchers to record and share research information, experimental data and protocols, and unpublished results. In this chapter, we first review previous research on OLNs, and then analyze the site Openlabnotebooks.org. We aim to illustrate the potentials of the genre for open sharing and to explore the communicative purposes of OLNs and the discourse features that contribute to accomplishing these purposes.

5.2 Open Laboratory Notebooks: Tools for Collaborative Knowledge Production

OLNs are tools with which researchers, mainly from chemistry and biology, where experimentation plays an important role, document their research processes online so that they are freely visible to others. Bradley (2006: online) coined the term 'open notebook science' to refer to 'a way of doing science in which – as best as you can – you make all your research freely available to the public and in real time', and emphasized that it is necessary for everybody to have access to all the information used by researchers to draw their conclusions. OLNs are used to report all the steps and data of ongoing laboratory research, not only the positive results of the research. Researchers also document their attempts to get a result, failed experiments, the challenges and problems they face and unpublished experimental data.

Open notebook science is part of the open science movement which, as explained in Chapter 3, has led to a variety of practices to make science more accessible and transparent. A common practice in some disciplines (e.g. physics/astrophysics) is the sharing of pre-prints by posting them to websites, pre-print servers or repositories such as ArXiv and SSRN, which help researchers to stake a claim to their findings or obtain informal feedback (Acord & Harley, 2012). Open notebook science goes beyond the sharing of pre-prints; it is closely linked with the democratic school of open science, which is concerned with open access to data and results free of charge, and with the transparency of the research process. However, researchers may be reluctant to share their data for several reasons, for example, the wish to produce more publications using these data, the fear of being scooped in attribution and thus not receiving due credit, concerns about having their research practices scrutinized or about sharing ideas that may not be accurate, or lack of time to curate the data and notetaking (Acord & Harley, 2012). To counter this reluctance, some authors have promoted the benefits of practicing open science and using OLNs, such as collaboration, efficiency and productivity. Dawson (2012) and Schapira and Harding (2019) argue that, by facilitating collaboration, open science (and OLNs) contributes to an increase in the speed and quality of scientific discoveries and advances: researchers can form partnerships and, if data are accessible during experiments (instead of keeping them secret until results have been published), they can build upon others' research, which may help to reduce the unnecessary repetition of experiments. Other researchers can reproduce the detailed protocols included in OLNs and enhance the quality of the results. Researchers can follow others' work, contribute comments in OLNs and thus provide insights and identify anomalies or weaknesses. Schapira and Harding (2019) add that other researchers can also gain important insights from negative data, which are rarely provided in traditional genres of publishing. They also refute the belief that openly sharing one's research online may lead to being scooped before publishing the results in a peer-reviewed journal by arguing that publishing in OLNs helps to claim temporal priority for the data, as results in OLNs are date-stamped. Additionally, OLNs may help researchers to get funding and grants and start new collaborations with other researchers.

However, even if, as noted by Wickman (2016: 13), open notebooks contribute to developing open access 'as a viable paradigm for scientific communication', writing them remains an activity with few practitioners and perhaps this is why there has been little research on the OLN as a genre and on its relationship with its printed counterpart. Clinio and Albagli (2017: para. 9) claim that the OLN is not 'an incremental improvement' on traditional laboratory notebooks, but a new way of doing and communicating science that provides access to all research records to promote collaborative production of knowledge, improve the circulation of scientific information and foster a type of peer review based on the quality of evidence, rather than on the authority of the experts chosen by journals. Clinio and Albagli's description of OLNs is based on Bradley's (2007, cited in Clinio & Albagli, 2017) open lab. The UsefulChem Project, Bradley's initiative to make the scientific research conducted in his laboratory as transparent as possible by publishing the research process in real time, adopted a 'bliki' configuration, consisting of (i) a wiki (the UsefulChem wiki), which made it possible to record the different steps and changes in the experiment; and (ii) blogs, used for project discussion (UsefulChem blog), for the recording of laboratory experiments (UsefulChem Experiments blog) or as a database of molecules (UsefulChem Molecules blog). The wiki page where the experiment was reported had a prescribed structured format with nine sections (number of experiments, graphical representation, researcher, objective, procedures, results, discussion, conclusion and log), the most important of which, Bradley claims, was the log, in which an objective and precise style was used to record the experiment while it was being conducted (Bradley, 2007, as cited in Clinio & Albagli, 2017). This prescribed structure highlights the value of partial results.

Comparing the models of communicating science used in these two genres, Clinio and Albagli (2017) discuss the difference in the rhetorical work involved in writing open lab notebooks and scientific articles. When writing research articles, researchers need to create a coherent narrative where results are presented as ready science (as opposed to science in construction), in order to persuade the readers of the discovery of a scientific fact. By contrast, open lab notebooks present 'storyless experiments' (Bradley, 2007, as cited in Clinio & Albagli, 2017); the focus is on science in the making and therefore both partial results (ongoing, finished, discarded) and final results (positive and negative) are important. Thus, OLNs foster a new epistemic culture, where greater value is placed on the detailed documentation of research practice and on open collaboration and discussion of the research process than on the construction of scientific facts. This is in line with Wickman's (2016: 17–18) observation that laboratory notebooks are used by researchers to generate knowledge and interpret results in the flow of laboratory research and to share their data, thus promoting a change in the scientific research culture, in which collective knowledge production is valued.

Wickman (2016) analyzed an open notebook science site, Open Wetware, used by biologists to post protocol and research information, which can be edited by anyone with access to the site. He characterized OLNs as dynamic rhetorical spaces 'within which groups of scientists and other stakeholders can present, discuss, revise, and circulate information' (Wickman, 2016: 12). This characterization distinguishes them from other types of notebooks and raises the issue of the boundaries between genres that have evolved from printed counterparts and new genres in online environments (see Chapter 2). Wickman (2016) points out that open notebooks are different from printed notebooks at least in two ways: their entries are designed for public use and can be modified by users other than the original writer.[1] Thus, he adds, it is difficult to determine whether open notebooks are 'an extension of the genre or, more radically, a complete break from it. What we do know is that the rhetorical situation for note making is changing as it begins to accommodate new media, wider audiences, and novel exigencies related to open access' (Wickman, 2016: 16–17).

Understanding the OLN genre also requires understanding its discourse-pragmatic features and how and why they differ from those of the traditional paper lab notebook. To the best of our knowledge, only Carter-Thomas and Rowley-Jolivet (2017) have explored the pragmatics of this genre. In their analysis of the online notebook written by a young researcher in genomics, they focused on the inter-generic and intra-generic adaptation that takes place when (i) the experimental data and results recorded in the notebook are elaborated to write a research article; (ii) the traditional paper notebook migrates to the web. From their comparison of the notebook and a research article (whose first author was the author of the notebook), Carter-Thomas and Rowley-Jolivet reported differences in the use of subject pronouns, author roles and verb forms. For instance, the personal pronoun 'I' is highly frequent in the notebook, reflecting a personal style in the description of the experiment. Verbs used to construct an argument or to evaluate others' claims (e.g. 'demonstrate') are frequent in research articles, but rare in notebooks; cognition verbs, used to conceptualize the research, are much more frequent. Their comparison of the online notebook with a corpus of 15 paper notebooks also revealed differences in style and presentation. While the style of the paper notebook is telegraphic, with a high degree of ellipsis, in the online notebook ellipsis and non-standard abbreviations are avoided, reflecting that the relation between the writer and the audience has changed from one-to-few to one-to-many. In addition, in online notebooks the language is informal, with non-standard grammatical patterns. Emotive elements, such as attitudinal adverbs or evaluative lexis and typographical signs typical of computer-mediated communication (CMC) (e.g. frequent use of punctuation marks, emoticons) are frequent, which Carter-Thomas and Rowley-Jolivet attribute to the need to connect with the audience.

The differences in discourse features between the research article and the OLN seem to reflect the distinction between what Gilbert and Mulkay (1984) identify as two interpretative repertoires used by scientists when describing their work: the empiricist and the contingent repertoires. The empiricist repertoire, typically employed in research articles, is characterized by an impersonal and detached style (e.g. avoidance of 'we' and of human agents) and by linguistic features that present scientific knowledge as objective and deriving unproblematically from empirical research. By contrast, the contingent repertoire is deployed in informal discussions of research and reflects the researchers' acknowledgement of the contingent nature of research: it is used to depict activities in the lab as shaped by personal inclinations and social factors. Traditionally, the contingent repertoire has been restricted to face-to-face discussions of the contingencies of everyday laboratory practices. However, Hine (2002) found that some aspects of laboratory talk were transferred to the online setting when scientists use newsgroup discussions to talk about local concerns of laboratory activity. As open lab notebooks are intended to share and discuss laboratory data and protocols, and could therefore be considered online 'laboratory talk', it is not surprising that they would display the features of the contingent repertoire.

5.3 Research Case Study: An Analysis of OLNs on the openlabnotebooks.org Site

This study explores openlabnotebooks.org, a site launched in 2018 by researchers at the Structural Genomics Consortium (SGC) to facilitate the online sharing of results and data in real time. The purpose and rationale of the site are explained on the openlabnotebooks.org home page, providing insights into the rhetorical exigence of the open lab notebook genre:

> We believe that making our research, data, and protocols available on a day-to-day basis will generate scientific ideas and discussions, avoid redundancy, foster collaborations, and accelerate progress.

In addition to enhancing the impact of their research, the purpose of the researchers who write these notebooks is to contribute to the development of the open science movement in the life sciences (Schapira & Harding, 2019: 5). For this purpose, the site openlabnotebooks.org takes the form of a blog (see Figure 5.1) where researchers working at various laboratories write posts about their experiments, explaining the motivation of the experiment and reporting on the research process (steps,

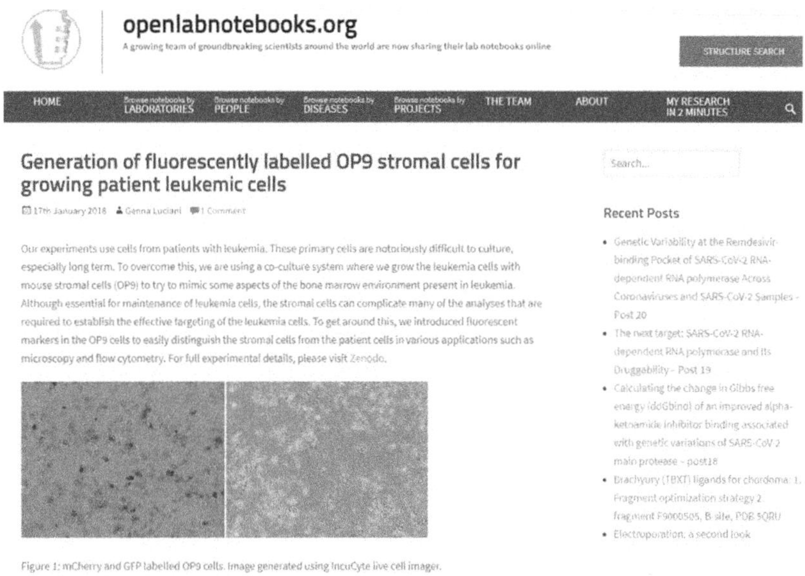

Figure 5.1 Post in openlabnotebooks.org (https://openlabnotebooks.org/generation-of-fluorescently-labelled-op9-stromal-cells-for-growing-patient-leukemic-cells/) by Genna Luciani. Licensed under Creative Commons Attribution 4.0 International

difficulties, positive and negative results). The posts written by the different contributing researchers appear in reverse chronological order in the blog (see Figure 5.2). However, all the posts by a single researcher occur together on the researcher's page (see Figure 5.3) and can be accessed through the link 'browse notebooks by people' on the navigation bar. All these posts make up the researcher's lab notebook, where readers get a more complete narrative of his/her research.

Schapira and Harding (2019: 4), two of the researchers who launched the site, describe the blogposts as being written in non-technical terms, which can attract collaboration with researchers from complementary disciplines. A post typically links to a much more technical document uploaded to Zenodo (https://zenodo.org) (an open data repository which assigns digital object identifiers [DOIs] to each file uploaded, thus supporting data attribution and citation), which provides a detailed record of the experiment, including all data and protocols (Figure 5.4). The blog is open to comments and therefore readers can comment on the experiment, provide feedback and ask questions.

The aim of the study is to analyze a small corpus of posts in OLNs (hosted on openlabnotebooks.org) and determine how their rhetorical and linguistic features contribute to and are shaped by the purpose of the OLN genre: to share information about the activity carried out in the lab in order to promote collaboration in knowledge construction. We

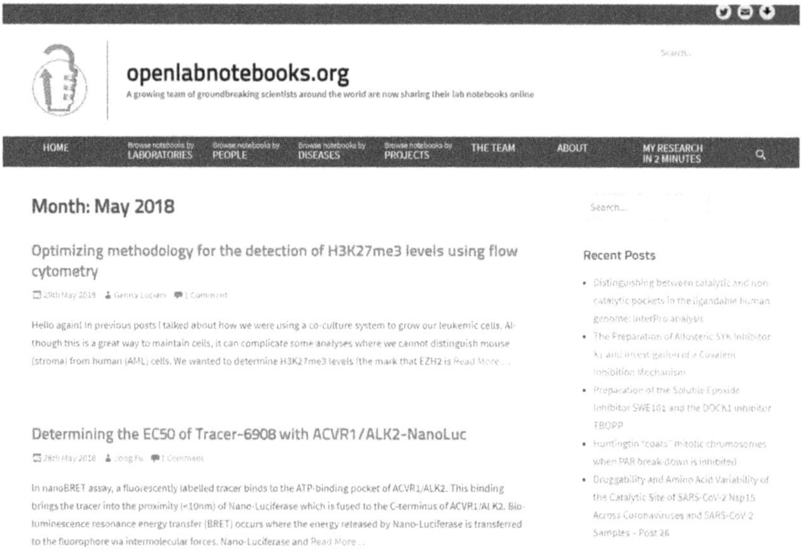

Figure 5.2 Screenshot from openlabnotebooks.org (https://openlabnotebooks.org/2018/05/). Licensed under Creative Commons Attribution 4.0 International

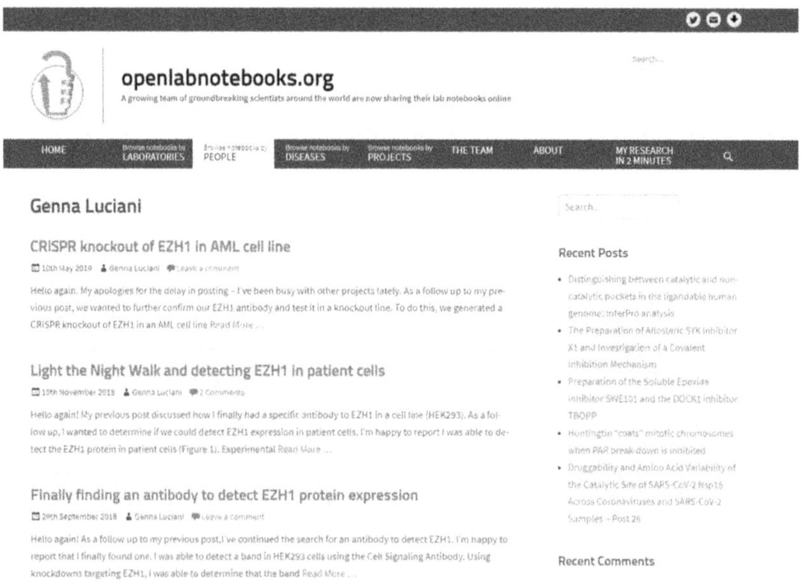

Figure 5.3 Screenshot from Genna Luciani's notebook on openlabnotebooks.org (https://openlabnotebooks.org/category/past-scientist/genna-luciani/). Licensed under Creative Commons Attribution 4.0 International

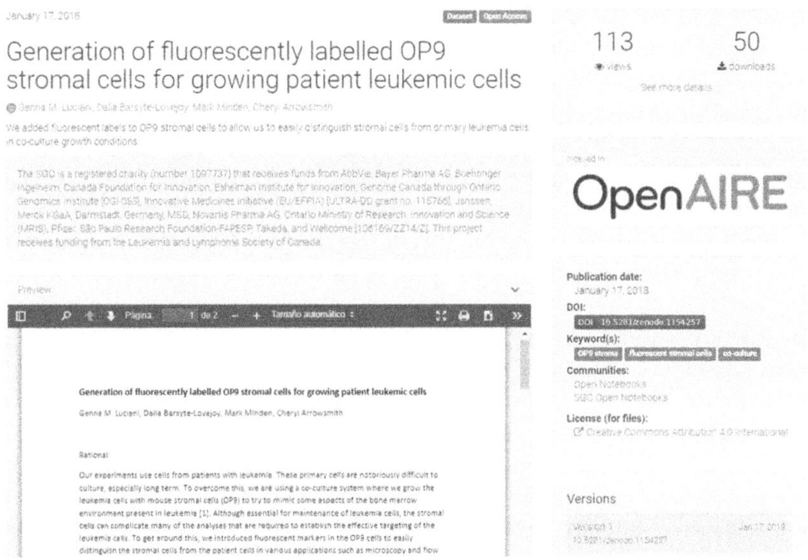

Figure 5.4 Technical description of the experiment in Zenodo (https://zenodo.org/record/1154257#.X33H_8IzaUI) by Luciani, Barsyte-Lovejoy, Minden and Arrowsmith. Licensed under Creative Commons Attribution 4.0 International

address the following questions: (i) What do scientists do with this genre and how do they harness the affordances of the medium for networked participatory practices? (ii) How do researchers use language to talk about lab activity online and engage in collaboration?

Data consisted of posts posted on the website from January 1, 2018 to January 31, 2020. For months when more than 10 posts were published, only the newest 10 posts were selected. The corpus is composed of 209 posts, written by 38 researchers from 13 laboratories (six in Canada, two in the United States, three in the United Kingdom, one in France, and one in Sweden). When the posts attracted comments, we also collected these for analysis. All the posts and comments were written in English, which is the language used by all the researchers to share their notebooks through the platform openlabnotebooks.org. Actually, among the challenges of creating open lab notebooks, Schapira and Harding (2019) mention the language barrier that may exist for users of English as an additional language.

5.3.1 Analysis of the posts

A first reading reveals that, while all the posts are concerned with the process of research in the laboratory, there is considerable variety. A few of them give general information about the writer's research, describing the purpose and giving an overview of the method, or providing background on the object of the research and its relevance. These are always 'first posts', which provide the context for the presentation of data and protocols in later posts (e.g. 'My first post here will be a brief introduction. More in-depth experiment and data posts are coming soon!'). A few posts are also employed by the writers to present their research so far, stating the purpose of the experiments carried out and the results. However, most of the posts are used for one (or both) of these two purposes: to describe specific methods and protocols and/or to report on results, whether positive or negative, after following a specific protocol.

The first stage of our analysis consisted in identifying moves and steps in the posts included in the corpus. In genre analysis, 'move' refers to 'a discoursal or rhetorical unit that performs a coherent communicative function in a written or spoken discourse' (Swales, 2004: 228) while 'steps' are lower-level functional units: 'elements that, together, or in some combination, realize the move' (Biber *et al.*, 2007: 24). In order to identify moves and steps, the data were analyzed with the Atlas.ti qualitative analysis program using an inductive approach. In other words, we generated coding categories on the basis of our observations of the data. We further analyzed: (i) the linguistic realization of these moves/steps; and (ii) the types of links included in the posts, to determine how these links contribute to the purpose of the genre. As pointed out above, we also analyzed the comments to the posts. These comments tend to be quite short, each of them performing a single communicative function. Therefore, they were first coded in terms of function, to establish why other researchers comment on the posts and therefore how they collaborate in knowledge construction. Secondly, the linguistic features of the comments were explored.

Researchers contributing to openlabnotebooks.org write posts about research in progress, and thus the content of the posts depends on what the researcher wants to report on at a specific moment during the experiment (e.g. description of a protocol, failure or success in an experiment). Therefore, given the heterogeneity of the posts, we did not find a common rhetorical structure in terms of moves. We found the following moves, although none of them appears to be compulsory: Move 1: setting the context; Move 2: describing and evaluating methods and protocols; Move 3: presenting and explaining results of an experiment; Move 4: expressing feelings related to one's own work as a researcher; Move 5: addressing the reader to Zenodo; Move 6: closing. Table 5.1 shows the different moves (and steps) that occurred in OLN posts and their frequency of occurrence.

Most posts begin by setting the context in one of various ways (Move 1). Writers may begin by presenting the purpose of their research or, most often, the purpose of the specific experiment or protocol described in the post (Step 1); describing the object of the research or defining

Table 5.1 Moves and steps in the posts in openlabnotebooks.org

Moves	Steps	Examples	Number of posts	Posts including this step (%)
Move 1. Setting the context	Step 1. Presenting the purpose (of the whole research, of a specific experiment)	My goal is to develop small molecules that inhibit USP5 selectively.	101	48.3
	Step 2. Describing the object of the research or defining terms and concepts	ALK2 and ALK5 are Type I receptors on the cell surface that can be activated by specific ligands.	42	20.1
	Step 3. Presenting background information important to understand the post	Two key features of the structure of HAO1 are important for the understanding of our data: the gating loop (figure 1A) and…	48	23
	Step 4. Referring back (to a previous post; to a previous step in the research)	In my previous post, I used baculovirus to overexpress HTT in mammalian cells.	79	37.8
	Step 5. Informing of activity between posts	It's been a little while since I posted (…) I've been doing some processing of…	31	14.8
Move 2. Describing and evaluating methods and protocols	Step 1. Describing steps of a protocol	Eight iterations later, I now have an initial proof-of-concept dataset, and a (mostly) final protocol. The general scheme of the assay is as follows: Coat the plates with laminin, a protein that allows the cells to stick to the plate, instead of growing as floating spheres (which is the current growth mode of choice in the field) (…) After four days…	95	45.4
	Step 2. Evaluating a particular method, technique, protocol or course of action	Now I know that treating cells with PEI at a ratio of 4x the volume of PEI to the mass of DNA transfects more cells than the previous method…	32	15.3
	Step 3. Justifying choices (regarding techniques/protocols)	To compare the two methods, I performed them both on the cells with a plasmid that will express enhanced green fluorescent protein (EGFP), because it visibly marks the cells which have been transfected.	72	34.4
	Step 4. Giving advice on how to solve problems that may arise at a specific step	Work-up and purification of these cores gave a bit of trouble due to the poor solubility in organic solvents. But luckily there is a trick that helps! [*explanation of the trick*] (Picture 1).	22	10.5
	Step 5. Describing the following steps in research	Within a month (hopefully) I'll be convincing C2C12 myoblast cells to express all kinds of mutant ALK2 to test its activity and interactions with other proteins within the cell. I'll be doing this by treating them so that …	73	34.9

(Continued)

Table 5.1 (*Continued*)

Moves	Steps	Examples	Number of posts	Posts including this step (%)
Move 3. Presenting and explaining results of an experiment	Step 1. Stating the results of an experiment	PRMT5 inhibition with GSK591 or LLY283 led to the decrease in SDMA.	105	50.2
	Step 2. Informing of the success or failure of a protocol/step/ course of action	I was successful in purifying various USP5 constructs with a biotin tag... Unfortunately, this construct did not generate infectious particles relative to empty-vector or GFP...	108	51.7
	Step 3. Interpreting results, presenting hypotheses and possible explanations	In cells overexpressing NSD3short, I observed greater eccentricities in their appearance with some cells displaying long projections. This may be indicative of an epithelial-mesenchymal transition (EMT).	94	45
Move 4. Expressing feelings related to one's own work as a researcher		...fingers crossed we get more crystals and more data! Yesterday turned into a litany of disappointment.	39	18.6
Move 5. Addressing the reader to Zenodo		For full experimental details, please visit Zenodo.	138	66
Move 6. Closing	Step 1. Asking for help	Anyone with expertise on this assay, could you please advise?	10	4.7
	Step 2. Inviting feedback or questions	If you have any questions or comments or want more detail on anything, (...) feel free to contact me.	23	11
	Step 3. Anticipating content of following post	In my next post, I will explain how...	34	16.2

specific concepts or terms (Step 2); or presenting a short piece of technical information important for understanding the post, or information to justify the relevance and need for the research protocol (Step 3); or they may begin by referring to a previous post or previous step in the research, which shows that the posts written by a researcher make up their lab notebook (Step 4). Some posts begin with a short report of what the writer has been doing (both in the lab and academically) since the last post they wrote (Step 5). The setting of the context tends to be short and quite personal (e.g. 'my goal', 'in my previous post'), as all posts, except for the first by each researcher, are part of a series of preceding posts.

Move 2, a description and/or evaluation of methods and protocols, occurs in many of the posts. In this move, the writer may provide a detailed step-by-step description of a protocol which has worked for them so that readers know exactly how to proceed if they want to do a similar experiment, thus helping replicability; or they may give a very brief outline of the steps followed because a more detailed description is presented in a hyperlinked document in Zenodo (Step 1). The example for Step 1 in Table 5.1 reflects a conception of scientific practice as an iterative process of trial and error (e.g. 'eight iterations later', 'proof-of-concept dataset', 'a (mostly) final protocol'). Methods, techniques or specific decisions at a particular moment in the research are evaluated and compared with others, often drawing on personal experience (Step 2). To increase reliability and trustworthiness, researchers' choices of techniques and protocols tend to be justified and explained (Step 3). The researchers also give advice on how to solve problems that may arise at a specific step (Step 4). They sometimes end a post where they describe protocols or where they present partial results by anticipating the following step (Step 5). As the examples of Move 2 in Table 5.1 show, the language used reflects the contingent nature of lab activity: the agency of the researchers is presented as important to get the expected results, which sometimes depends on knowing the tricks; cells need to be 'convinced' to do what is expected of them (see example for Step 5).

Move 3 consists of the presentation and explanation of the results of an experiment, protocol or a step in the experiment. This may involve presenting results or laboratory observations (very often accompanied by figures illustrating these observations) (Step 1); or reporting whether a protocol/experiment/step worked, did not work or worked only partially, and commenting on problems (Step 2). In most cases, the researcher also informs the reader of the consequences: they achieved (or did not achieve) the expected results and can move on to the next step or has to find another way to get what they need and thus repeat the experiment. As the following fragments of posts written by the same researcher at different moments (Examples 5.1 and 5.2) show, this step is typically realized through evaluation, with a high frequency of attitudinal adjectives and adverbs (e.g. 'sadly', 'I'm absolutely thrilled', 'It's amazing!').

Again, the discourse emphasizes the importance of human agency (see italicized fragments in Examples 5.1 and 5.2).

> **Example 5.1** I ended last time saying *Shubhash had saved the day by finding a better virus stock* and growing up some more cells for me. Sadly, this did not work – still no expression (Adamson, 11 Jul. 2018)

> **Example 5.2** So, the week has been super busy, but all for good reasons this time!! I'm absolutely thrilled to say that amongst all the salt crystals I found in my crystal plates, I also found the crystals below, one of which diffracted to 1.25 Å (…). It's amazing! You can see everything – as shown in the pic below (…) *Clearly the use of the composite screen I set up has increased the hit rate* (Adamson, 19 Oct. 2018)

The researchers also usually interpret the results, present hypotheses, provide possible explanations for their observations and positive results or reflect on the reasons something went wrong (Step 3), which may lead them to do things differently. This step is realized through expressions of epistemic stance to put forward explanations tentatively. In order to express tentativeness when providing explanations, writers' statements are usually hedged with expressions of possibility (e.g. 'It appears that', 'this may be due to', 'this could be because', 'it is possible that', 'there are several reasons') (see the example for Move 3, Step 3 in Table 5.1).

In some of the posts, the writers express their feelings (disappointment, excitement, happiness, nervousness) not only about the progress of their experiment but also about their work as researchers in general (Move 4), using a personal style and a 'familiar' tone, typical of the contingent repertoire (see examples in Table 5.1). This sharing of feelings that readers may have also experienced helps to construct solidarity and thus promote collaboration.

A high percentage of posts (66%) refer the reader to detailed information about the experiment in Zenodo (Move 5). Although this move tends to occur at the end of the post, at times it occurs at the beginning before the researcher comments on the experiment in question.

Finally, posts may include a closing move (Move 6), where the poster asks for help (usually to understand why something happened), invites feedback or questions, or anticipates the content of their next post.

Although in this discussion we have focused on text, posts also include visuals (mostly microscope photos to convey observations and results) (see Figure 5.1) and links to different sites. Table 5.2 presents a list of the types of links present in these lab notebooks.

These links provide the information that the poster considers necessary for other researchers to understand the post, to understand the reasons for the writer's decisions and to repeat the protocols or techniques described in the post. As noted above, many of the posts include links to documents at Zenodo that provide a more technical and impersonal

Table 5.2 Types of links in openlabnotebooks.org

(1)	Links to technical documents at Zenodo.
(2)	Links to other output by the writers (e.g. posters, pre-prints).
(3)	Links to a previous post reporting on previous steps in the research.
(4)	Links to others' output or texts with useful information (e.g. papers describing the technique being used, papers with a review of literature useful for readers of the post, posts in another blog).
(5)	Links to pages providing information on products mentioned in the posts (e.g. Wikipedia, specialized pages about chemical substances).
(6)	Links to other data repositories or to free software (e.g. software to view data).

report of the protocol. Compared with the post, the report provides much more detailed and technical information to facilitate reproducibility. In addition, the report is impersonal, avoiding references to the researchers' agency. See Example 5.3, where a fragment of the technical report linked to a post is presented, and compare it with the example for Move 2, Step 1 in Table 5.1.

Example 5.3 For HSJD-DIPG-07 cells to grow adherently, 96-well tissue culture plates (Greiner cat. no. 655090) (recommended for use with Celigo image cytometer) must be coated with laminin (Engelbreth-Holm-Swarm murine sarcoma, Sigma, cat. no. L2020) as follows:

- Dilute laminin 1:100 in sterile water
- Add 45 μL of laminin to each well*

5.3.2 Analysis of the comments

The communicative functions of comments in posts can be grouped into three broad categories: requesting, expressing solidarity and moral support, and providing practical advice (Table 5.3). These functions reveal the collaborative nature of OLNs. The writers of the posts also write comments in response to comments by readers, engaging in cooperative talk.

As stated earlier, all these comments are characterized by being brief and going straight to the point. They read very much like off-the-cuff conversation that might take place in the lab. However, when asking for help or collaboration or when asking writers to justify their experimental choices, commenters tend to include a brief narrative about their research to contextualize the request (Example 5.4). These are face-threatening acts, and therefore these comments tend to include politeness expressions to minimize the threat (e.g. 'was wondering'). By contrast, questions about findings or observations are usually more direct (Example 5.5). The answers to these questions are often elaborate, including justifications of the researchers' choices or explanations of results.

Example 5.4 I am from Australia, we are also interested in using the BacMam system, we will be using CHO cells, was wondering if you could share a detailed protocol for the transfection and transduction.

Example 5.5 Very cool, Jolene. What PRMT5 inhibitor concentration did you use?

Some comments are intended to express solidarity and moral support. A few comments are phatic posts whose only purpose is to maintain the social relation between the writer and the commenter (Example 5.6). Other comments are used to thank the researcher for sharing, to express interest (e.g. Example 5.7), to evaluate the researchers' work positively and offer congratulations, or to show empathy (e.g. Example 5.8).

Example 5.6 Commenter: The crystals look kinda ugly :>

Example 5.7 Very useful for our current works in mathematical models of control malaria, as we can incorporate now this annual cyclicity.

Example 5.8 Hi Ros! Sorry to hear about your problems with protein expression.

The use of informal expressions ('kinda', 'sure'), abbreviations ('congrats'), emoticons and shortened first names (e.g. 'Ros') is very frequent in these comments. This constructs an intimacy that fosters collaboration. Comments such as Example 5.7, which acknowledges the usefulness of the information in the post for the commenter's own research, reveal how sharing through lab notebooks, instead of waiting for the research to be published, can help advance science.

At other times, comments are intended to provide practical advice and information, particularly when the writer has reported problems

Table 5.3 Communicative functions of comments

Comments by readers	Response by writers of the post
Requesting Asking for help and collaboration Asking questions about findings/results Asking the writers to justify experimental choices	Offering requested help Answering questions
Expressing solidarity and moral support Phatic talk Thanking for sharing Expressing interest Positive evaluation Showing empathy	Phatic talk
Providing practical advice Offering help Providing useful information or sharing knowledge of interest for writer of post Sharing similar results Suggesting areas of further research Suggesting ways to overcome negative results Sharing tricks to solve a problem	Thanking for suggestion Asking questions after suggestion Evaluating suggestions positively and accepting them Explaining why suggestions may not work
	(to various comments) Thanking for comment Sharing knowledge and resources

with the experiment described in the post. Commenters may offer help, provide links to information that may be useful, share similar results or suggest ideas for further research. Commenters may also suggest ways to overcome problems or negative results (Example 5.9a) and share tricks to solve these problems (Example 5.9b).

Example 5.9

(a) I'm sure you're already on it but perhaps you could try dehydrating them [crystals] or seeding them in a new plate screen? Has worked for me in the past. Best of luck!
(b) One of my protein crystals were also grown inside the membrenous drops, as you described. (...) I added 5-15% ethylene glycol in the protein condition & that solved the membrane problem.

When suggesting new actions, commenters tend to use politeness devices to make sure that the writer's competence is not challenged ('I'm sure you're already on it but perhaps you could try'). The contingent repertoire is clearly reflected in these comments, where the advice provided is often experience based (e.g. 'has worked for me', 'that solved the membrane problem'). As pointed out in Hine's (2002) study of newsgroup discussions on laboratory activity, experience-based responses in these comments point the author of the query towards some action that has been successful. OLNs facilitate the sharing of experienced-based knowledge and make it possible to learn from someone who masters the technique, rather than from published protocols (Hine, 2002: para. 4.19).

This case study suggests that the OLN is a new genre that harnesses the affordances of the web for open and fast sharing of research, responding to the exigences of open science. The rhetorical analysis has shown that scientists from different laboratories use this genre to share laboratory protocols, data and preliminary results (positive and negative) and to engage in practical collaborative discussion of laboratory activity. In addition to giving access to findings before they are published, this genre also provides access to information that does not typically have a place in formal genres of science communication (e.g. negative results, experience-based solutions and tricks) and serves purposes not achieved with other genres (e.g. networked discussion on specific laboratory problems). The analysis also shows that when writing OLNs, researchers use discourse features (e.g. informality, personal language, aspects of the contingent repertoire) that tend to convey a sense of solidarity and promote collaboration, and at the same time link to documents with detailed technical information, written in an impersonal style, to facilitate evaluation and reproducibility. Although individual posts seem to focus on specific issues or protocols of a scientist's research, the set of posts as a whole comprising each lab member's lab notebook usually provides a more comprehensive insight into their research, thus opening the laboratory doors and inviting continuous open peer feedback and discussion. In sum, this genre should be understood within a new paradigm of doing and communicating science. It is a genre for the sharing of contingent everyday laboratory practice and for collaborative knowledge production.

5.4 Concluding Remarks

In this chapter, we have discussed how a genre for making laboratory notes has migrated to the internet and evolved into a (new) genre for sharing research in progress with peers. This digital genre leverages the hypertextual and interactive affordances of the online medium to support knowledge production and communication. Thanks to these affordances, OLNs enable scholars to fulfil multiple rhetorical goals, namely, to facilitate the accountability, reliability and reproducibility of scholarly research and to strengthen collaboration and knowledge sharing for the advancement of science. The use of OLNs by teams of researchers from various laboratories discussed in this chapter supports Acord and Harley's (2012: 391) prediction that digital tools for discussion of research in progress will be adopted when they address 'distinct needs in a specific discipline'. These teams benefit from openly sharing data and partial results and collaborating to solve problems, while receiving credit for their research results by uploading them to Zenodo and thus making them citable.

The analysis of OLNs has important implications for genre theory, since it provides theoretical insights into digital genres. The posts in openlabnotebook.org can be regarded as what we call 'nested genres', that is, several instances of a genre (in this case posts), put together to form another genre (the OLN). All the posts written by a researcher compose a genre instantiation in progress: his/her open lab notebook. Although each post tends to have a more specific purpose (therefore, posts may differ considerably in the moves and steps they include), all of them contribute to the purpose of the OLN: to share research in progress in order to make it reusable by others and to get feedback.

In Chapter 6, we continue to focus on sharing and collaboration by exploring how academic social networking sites (ASNSs) are used for these purposes and discussing the role of the Q&A feature as a genre for intercultural informal interaction and peer support online.

Note

(1) Wickman (2016) adds the caveat that his definition of open lab notebooks differs from others (e.g. Bradley's [2007, as cited in Clinio & Albagli, 2017]) for whom 'open' means that notes are readily available online, but not that they can be modified by other users.

6 Interacting in Academic Social Networking Sites

6.1 Introduction

The affordances of digital media enable scholars to connect with other peers across the globe, share resources and establish new kinds of relationships. Social networking sites (SNSs) are web-based tools that allow users to construct a profile and build a network with other users with whom they share connections and interests (boyd & Ellison, 2007). Academics use a variety of SNSs, both sites aimed at the general public (e.g. Twitter, Facebook, LinkedIn) and those targeted mainly at scholars, that is, academic social networking sites (ASNSs) (e.g. ResearchGate [RG], Academia.edu). Jeng *et al.* (2015: 890) characterize ASNSs as social platforms that can 'help scholars to build their professional networks with other researchers and facilitate their various activities when conducting research'. The term covers a variety of online platforms developed to help academics harness the benefits of online networking by combining elements of social networks (e.g. the creation of user profiles within a bounded system, interaction with a list of users) with features and tools adjusted to the needs of academics, such as the uploading of articles and tracking of citations. Jordan (2019) distinguishes two types of ASNSs: those developed mainly for profile creation and networking (e.g. Academia.edu, RG) and those created as bibliographic reference managers, used originally to organize and share publications, which have added social networking features (e.g. Mendeley, Zotero and CiteULike). In this chapter, we focus on the first group. At the time of writing, the two most popular ASNSs in this group are Academia.edu and RG, which claim on their websites that they have 147 and 19 million registered users, respectively. These ASNSs also state that their mission is to help users share their research and enhance its visibility, connect with other researchers and keep up to date with the research of colleagues. RG also promotes itself as a place where researchers can ask questions, get answers from colleagues and solve problems.

In this chapter, we first present a brief review of ASNSs, focusing mainly on their affordances and researchers' concerns regarding their use, the reasons for their use and their role as sites for sharing, discussion and interaction. We then report on the results of a research case study which illustrates the use of the question and answer (Q&A) feature in RG as a forum for interacting with other researchers and a space for scholarly intercultural communication.

6.2 Academic Social Networking Sites

6.2.1 Affordances of ASNSs and researchers' concerns regarding their use

In their analysis of the potential of SNSs for communication and networking in academia, Nentwich and Konig (2014) describe the various functions developed by ASNSs to meet the needs of their users: user profiles, multiple communication channels within a single platform (e.g. discussion forums) and networking tools, such as contacts/followers, search functions, automatically generated requests to perform some actions involving other members or visualization of networks of co-authors. Drawing on previous literature on ASNSs, Meishar-Tal and Efrat Pieterse (2017) discuss the following affordances for researchers: management of an online persona; diffusion of research by uploading publications to the platform; cross-disciplinary and cross-border collaborations, and the possibility of establishing relations with other researchers (through tools like email and internal messaging systems); the collection and organization of academic information of interest to the user, such as publications by other researchers; and the measurement of impact, through a variety of social media metrics (e.g. number of citations, number of reads or downloads).

ASNSs can increase researchers' visibility within the network, which is facilitated by the 'following model' (Bonaiuti, 2015) used by these platforms: users can follow the activity of their connections, know when they add papers or ask/answer questions and interact with them. In addition, the opportunity to follow, view, download, bookmark, recommend or share the work of other researchers afforded by ASNSs facilitates the signaling of engagement and enables informal social reviews of that work (Chapman & Greenhow, 2019). Thanks to these affordances, ASNSs have gained popularity among scholars, who use them as repositories to make available their research outputs, as tools for networking and as forums to discuss research and academic-related issues and ask for advice, help, information or opinion (Barbour & Marshall, 2012; Bonaiuti, 2015; Thelwall & Kousha, 2015).

One of the main strengths of ASNSs is, therefore, the potential to support cross-disciplinary, transcultural communication and collaboration among researchers (Goodwin *et al.*, 2014). ASNSs could be characterized as what Leppänen and Peuronen (2012: 389) call 'translocal affinity spaces', places where users from different lingua-cultural backgrounds can share their interests with other like-minded people. The term 'affinity spaces' was coined by Gee (2004) to refer to spaces where individuals who have affinity for a topic can meet and communicate to discuss this topic within groups. More recently, Gee (as cited in Hudson *et al.*, 2015) has stressed the following features of affinity spaces: they are based on an affinity for something, not on the basis of identity markers such as race, class or gender; they are about collective intelligence; there is no gate-keeping; they are fluid, without well-defined boundaries; standards are internal and roles are flexible, with beginners and professionals sharing the space, and with individuals participating to different degrees. Jones and Hafner (2012) also emphasize the affordances of digital media for the creation of 'globalized online affinity spaces' which facilitate intercultural communication. Despite the diverse backgrounds of the participants, these spaces develop their own culture, including norms of interaction. English is often employed in ASNSs to facilitate communication among scholars with various linguistic backgrounds. The cross-linguistic virtual communities using English as a lingua franca (ELF) in these settings tend to appropriate the resources of the language and adapt it to suit the context (Luzón, 2018c). The potential of these spaces for intercultural communication does not imply, however, that all researchers have equal opportunities to participate, as we discuss in the following section.

Despite offering these new opportunities for networking and collaboration, ASNSs also have a dark side. As noted by researchers, although Academia.edu and RG are presented by their founders as sites designed to facilitate sharing and open access, they are not educationally affiliated companies but rather for-profit companies funded by venture capital, which seek to monetize the networks (Bond, 2017; Hall, 2015). Although this monetization is mostly done by trying to attract as many users as possible to the site and by targeted advertisements, Academia.edu has also implemented other methods of generating income (Jordan, 2019; Laakso *et al.*, 2017). Its 'premium feature' subscription facility includes a range of services that allow users easier access to documents and to obtain data analytics about who is reading and citing their papers (Bond, 2017). Academia.edu describes its business model as 'Freemium', combining 'free access to research for everyone' with 'paid premium capabilities to subscribers', who cover the costs of the website. Hall (2015), however, questions this model, by stating that it is based on the ability of these sites to explore the data flows generated by the researchers using the site and thus provide useful information for R&D institutions about the most impactful papers. Hall (2015: online) warns that, in effect, academics are working for these private for-profit companies for free 'by providing the aggregated input, data, and attention value'.

In addition to ethical issues related to the commercial nature of these repositories, other important concerns that may influence researchers' adoption of ASNSs are issues of privacy (i.e. researchers' activity on the platform is tracked), data security and copyright (i.e. researchers may be infringing the law by uploading their publications to these platforms). Penn (2018) includes these ASNSs, together with pirate sites such as Sci-Hub or LibGen, within what she calls 'black/guerilla' Open Access, or peer to peer (P2P) sharing of articles or other research output. Although this practice is supported on the grounds that copyright laws prevent knowledge sharing and production, it is, however, a matter of concern for researchers, who need to check whether a publication can be legally shared before uploading it to these platforms. Hosting copyrighted material is a way of sharing publications that is opposed by academic publishers, who have tried to stop this practice by issuing takedown notices or calling for action against these ASNSs (Jordan, 2019; Penn, 2018).

6.2.2 Motivations for their use and uptake

As ASNSs offer such a variety of communication and networking possibilities for academics, studies have sought to understand why they are used by specific groups of researchers, and revealed these motivations:

(1) Showcasing and disseminating one's research output more widely and thus enhancing one's citations (e.g. Elsayed, 2016; Nández & Borrego, 2013; Veletsianos, 2012).
(2) Promoting one's CV, being more visible and more easily contactable (Menendez *et al.*, 2012; Thelwall & Kousha, 2015). Many respondents in Jordan's (2014) study indicated that their Academia.edu profile functioned like an online business card. Profiles make available to other members specific information selected by the user (Nentwich & Konig, 2014).
(3) Constructing an academic identity (Jordan, 2019; Thelwall & Kousha, 2015; Veletsianos, 2012).
(4) Discovering new research, being kept updated on publications of interest by following other academics, being notified when new research is uploaded and having access to publications (Manco-Vega, 2017; Nández & Borrego, 2013; Segado-Boj *et al.*, 2019).
(5) Establishing and maintaining contact with other researchers (for community building or collaboration) (Nández & Borrego, 2013).
(6) Interacting with peers and seeking advice and information from them (Goodwin *et al.*, 2014; Jeng *et al.*, 2015).
(7) Tracking the metrics of one's publications, such as the number of views and downloads or the university and discipline of the readers (Laakso *et al.*, 2017).

Motivations to use ASNSs appear to differ by user group (e.g. students, faculty members) and discipline (e.g. researchers in humanities vs. researchers in natural science or medicine) (Jordan, 2019; Thelwall &

Kousha, 2015), and by researchers' institutional affiliations (geographical location). Many studies on ASNSs have analyzed their uptake by researchers from diverse linguistic/cultural backgrounds, such as Arab scholars (Elsayed, 2016), Brazilian scholars (Manco-Vega, 2017), Spanish scholars (Nández & Borrego, 2013; Segado-Boj *et al.*, 2019), Israeli scholars (Meishar-Tal & Pieterse, 2017), Finnish scholars (Laakso *et al.*, 2017) and Italian scholars (Manca & Ranieri, 2016). Overall, these studies show that although sharing publications and enhancing citations seem to be a high priority for researchers, there are differences among researchers from different countries in the degree to which they use these social networks and their motivations for doing so. Menendez *et al.* (2012) analyzed data from 30,428 profiles on Academia.edu and found statistically significant differences based on the economic development of different countries in all the items that were analyzed (e.g. personal information shared, content contributed, popularity, number of questions asked, answered and followed, level of engagement), with users in highly developed countries scoring higher on all variables. Menendez *et al.* (2012) tentatively conclude that, despite the potential of ASNSs to provide equal opportunity for all researchers, Academia.edu seems to preserve the hierarchical geographical structure of academia, with characteristics such as university ranking or country development affecting visibility and participation.

Manco-Vega (2017) analyzed the motivations for using ASNSs by early career researchers and PhD students in the social sciences in Brazil. She discusses several motivations for using ASNSs (communication with peers, keeping abreast of research publications in their fields and having access to publication, dissemination of research, self-branding and participating in interest groups), the main perceived benefit being gaining more access to information. Through these sites, researchers can access not only pre-prints and unpublished papers, but also published papers not available in their libraries and which they cannot afford through payment to traditional journals. These sites are therefore particularly useful for scholars from developing countries since they increase their accessibility to published research.

Thelwall and Kousha (2015) also found differences between countries in the uptake of RG. The countries with the highest and the smallest number of publications, when compared with their publications in WoS-indexed journals, were, respectively, the United States and China, which reveals that US researchers benefit from this ASNS to advertise their research. This study also shows that some countries (e.g. Brazil and India) take more advantage of this social network than others (e.g. China and Russia), which could be related to different levels of familiarity with the use of social networks for research or to the country's academic culture. Some scholars have associated the practices in the use of ASNSs with the academic system of the country concerned. For example, Segado-Boj *et al.* (2019) found that the Spanish academics they interviewed used SNSs mainly to manage content, search for experts in their field, find new studies and learn more about their authors. They also suggest that one of the reasons why Spanish researchers do not fully exploit the potential of SNSs is the slow pace at which the institutions and organizations they deal with are adopting these tools.

Proficiency in English, access to technology and digital literacy may also be important factors influencing the adoption of ASNSs and limiting their use by some scholars. Manco-Vega's (2017) research also confirms that English is the primary language for communicating efficiently in these social media. She claims that the linguistic and cultural barriers found in the traditional channels for scholarly communication are reproduced in the digital context. Research on scholarly networking by Ibero-American researchers also points to linguistic barriers. Raffaghelli and Manca (2018) suggest that a lack of linguistic confidence to communicate in English may partly account for the low levels of interaction of most of these researchers with international colleagues. They note that the language in which science is produced in Latin America, Spain and Portugal may hinder the participation of these scholars in ASNSs. Segado-Boj *et al.* (2019) note that lack of technical expertise is also a cause for the slow adoption of ASNSs in Spain.

6.2.3 Collaboration and interaction

ASNSs may facilitate knowledge production and collaboration by making it possible for scholars to share research outputs, find (and be found by) new collaborators and harness the expertise and knowledge of a wide network of researchers. Two features of ASNSs that contribute to making them sources of scientific knowledge are document sharing and Q&A forums. Regarding the former, as pointed out above, ASNSs act as repositories that provide open access to researchers' output (including copyrighted publications). The latter feature allows researchers to draw on the knowledge of others. RG hosts a genre that is mainly used for multi-party interaction, the scholarly Q&A discussion. This function enables scholars to find answers to specific problems in their research, to request help and support to overcome specific difficulties or to seek advice on how to act in specific situations. This 'crowdsourcing' functionality, i.e. the ability to find answers to problems or questions by tapping into the expertise of the community (Bonaiuti, 2015) is illustrated by the case of Nicole Forster, a researcher on cancer treatment. When neither Forster nor her colleagues in the laboratory knew how to proceed in one of the steps of an experiment, she found the answer in one of the replies that she received using RG and saved time not trying to find the answer elsewhere (see Bartling & Friesike, 2014).

Some studies on ASNSs have focused on this latter feature of RG because of its potential for the open, informal exchange of information between peers across the globe. The Q&A platform is an open space where participation by new users is welcomed: anybody registered on the platform can use the Q&A feature to ask and reply to questions (Goodwin *et al.*, 2014). Users have the ability not only to discuss research-related factual information, but also to share opinions, seek advice and feedback and provide emotional support (Jeng *et al.*, 2015). To encourage participation, questions are tagged with topics (e.g. pragmatics, English for specific purposes, science communication) which correspond to tags for 'skills and expertise' of RG members. Members are sent notifications about questions to which they may contribute because they match their self-declared interests and skills.

Research about the RG Q&A has focused mainly on examining how scholars exchange information across disciplines (i.e. the types of questions they pose and the content of the responses) (Jeng *et al.*, 2017) and on evaluating the quality of the answers (Li *et al.*, 2018). These studies provide insights into the communicative purpose of the threads and on the rhetorical exigence of the Q&A genre. Jeng *et al.* (2017), for instance, analyzed threads (i.e. initial post, or 'question', plus responses) in three disciplines and categorized questions into information-seeking (e.g. 'Does anyone know about the sizes of remnant debris of comet Ison?') and discussion-seeking (e.g. 'In your opinion and experience which one is the better way that students could learn and enjoy History?'). Both questioners and respondents used their posts to add factual information, provide resources, refer to other researchers or theories, provide opinions and feedback and provide personal opinion. Similarly, Deng *et al.* (2018) analyzed a corpus of questions and answers in the field of Library and Information Science to determine the kind of questions that scholars ask on this topic and the features of the answers. They found the following three types of questions in order of frequency: asking for information (43.9%), opinion (36.9%) and suggestions (19.2%). A high percentage of the questions asking for suggestions or information included a description of the questioner's experience, making them more personal.

Although the possibilities for intercultural discussion offered by the Q&A feature in RG may lead to new communicative behaviors and social practices, research on these new scholarly practices is scarce. In particular, Q&A discussions in RG appear to be ELF settings, with English being the most widely used code for communication among the members of these multilingual groups. To date, though, there has been little research on the use of English in this forum. One exception is our study (Luzón, 2018c), which analyzed ELF interactions in RG Q&A forums to determine the extent to which participants in these discussions use strategies to achieve successful intercultural communication. The analysis of posts taken from 30 Q&A threads in RG (all with participants from different lingua cultures), including the initial post or question and the responses, reveals that participants make frequent use of four pragmatic strategies to ensure mutual understanding, signal solidarity, maintain rapport with interlocutors and make the conversation flow: (i) backchanneling; (ii) the 'let-it-pass' and 'make it normal' strategies (Firth, 1996); (iii) strategies to manage and respond to impolite behavior; and (iv) code-switching. The frequent use of backchanneling in the form of expressions of thanks, congratulation, approval and interest reflects the willingness to collaborate in knowledge construction of most interactants. A relevant feature of these interactions is the adoption of the 'let it pass' and 'make it normal' strategies (i.e. letting unclear utterances or non-standard forms pass), particularly if we consider the common belief in the need to adhere to standard English in other traditional written genres of scholarly communication. Many responses in the threads included instances of non-standard forms (e.g. 'If there will be publish negative results', 'I found interesting what follows, so I led you know it', 'This has nothing to see with'), but participants in the discussions prioritized content over standard linguistic usage and they made no reference to these forms or to any communication problems they may cause. In sum, the online medium represents a site for ELF users' interaction in which the use of non-standard forms does not necessarily impede exchange of knowledge, ideas, interests and motivations among the participants.

6.3 Research Case Study: ResearchGate Q&A as a Digital Genre for Intercultural Informal Interaction

In this study, we analyze RG Q&A threads where scholars discuss the issue of academic publishing in order to illustrate how these spaces act as forums for informal interaction between researchers from different linguacultural backgrounds and disciplines. Unlike other genres discussed in this book, which are used by scholars to disseminate knowledge, the RG Q&A is a forum to obtain peer support online, to seek information, help and advice, and to share concerns publicly with other scholars.

The data for this study consist of 426 posts taken from 12 Q&A threads in RG posted between 2014 and 2018, including the initial post or question and the responses.[1] The selected Q&A threads were related to the topic of academic publishing in English and the participants were researchers in different disciplines. Looking at the profile of the participants (e.g. their names and institutions), most of the contributors appeared to be multilingual scholars, mainly coming from the periphery or semiperiphery (e.g. India, Afghanistan, Malaysia, Iraq, Sudan, Egypt, Macedonia, Bosnia and Herzegovina, Estonia, Poland, Russia, Argentina, Colombia, Brazil, Spain, Portugal), although some posts were written by scholars from Northern Europe and by apparently 'native speakers'[2] of English. Some of the participants express the linguistic position from which they contribute (e.g. 'We … Spanish speakers', 'As a native speaker of English'), categorizing themselves into a particular sociolinguistic group, as a way to strengthen their opinions and/or to describe their linguacultural background.

The posts that triggered the thread were 'discussion seeking' questions (Jeng *et al.*, 2017), that is, questions to which there may not be a single answer, but where responses help to shed light on the question, or serve to share opinions about issues related to the topic. While some of these initial posts included only the questions, most of them also included some justification for the question (e.g. a description of the questioner's related experience) or the questioner's opinion or view on the issue. In most cases, the questioner requested advice or sought to trigger the sharing of experiences (e.g. 'How to handle reviewer's comment on grammatical mistakes?', 'Do you think that academic journals in general are fair when publishing articles by non-native English scientists?'). The length of the threads ranged from 10 to 105 posts, and most of the participants contributed more than one post to the discussion.

The posts in the corpus were analyzed for the kind of discourse functions they served, in order to study what scholars 'do' in these forums and the kind of interactions in which they engage. The data were imported into Atlas.ti 8 for coding and analysis using an inductive approach. As RG Q&A discussions have a collaborative purpose, we drew on research on cooperative online interactions (e.g. Ädel, 2011; Schallert *et al.*, 2009) to inform the coding process and interpret the results of the analysis. However, the list of discourse functions is based on our observation of the data. The analysis resulted in our identification of the discourse functions in Table 6.1. Some of the posts served only one function (e.g. *Agreeing*), while others included several communicative acts and were used by scholars to perform more than one function (e.g. *Agreeing* and *Giving advice*).

In line with our characterization of RG as an affinity space, the posts in the Q&A threads analyzed here are used to perform acts that reveal the interpersonal and supportive nature of these threads. Participants with different degrees of expertise ask for information, help and suggestions, share their concerns, opinions and experiences, and offer information, help and advice. *Asking for information*, *Asking for opinion* and *Asking for advice* are functions commonly performed in the questions that trigger the discussion, although some participants also contributed by asking for additional information, opinion or advice related to the issue being discussed (e.g. 'I'd liked to complete with another question'). Most of the posts are intended to provide the questioner with what they requested (*Providing information*, *Expressing opinion*, *Giving advice*), which is an indication of the collaborative nature of the threads.

Participants *provide information* by sharing knowledge or resources (links, embedded publications) that may be useful for the questioner.

Table 6.1 Discourse functions in the RG Q&A threads

Function	Examples
Asking for information	What features of English writing differentiate natives from non-natives?
Asking for opinion	Do you think that academic journals in general are fair when publishing articles by non-native English scientists?
Asking for advice	What can be done for language editing from a native speaker? (...) Please suggest!!!;[a] How I can handle this reviewer's comment?
Greeting	Hi, all; Dear+ *name*; Dear Dr+ *name or surname*
Closing	Best wishes; Have a great day!; I wish you good lucks!; Good luck, dear colleague!
Providing information	I am sharing some sites which may be useful for editing documents:
Expressing opinion	Yes, I think culture should be taken into account while sending papers for review.
Giving advice	I suggest that you have your manuscript reviewed by a native speaker from your specialty.
Offering help	I would be happy to edit a reasonable length journal article for a non-native speaker at no charge.
Expressing feelings and emotions	I'm amazed to see how...; this is fascinating.
Evaluating others' contributions positively	A worth million question it is; that's a very, very good question, dear *name*; congratulation for this smart view.
Thanking	Thank you for asking the question; Thank you all for your valuable comments.
Responding to thanks	You're welcome, *name*. :)
Asking for clarification	I am missing something here....are you looking at a manuscript submitted to a journal?
Agreeing	Yes Dear *name*, that's true. Very little percentage (in my country) who reads an English research; You are right. A lot of work needs to be done in other languages also; I agree completely with your say; Agree with *name*; I agree with others that the ideal course would be to...
Disagreeing (mild or mitigated disagreement)	Although I understand and respect the opinions of all our fellow scientists (...) I believe that we need an international language, English or not.
Recommending	*[done by clicking on 'Recommend' at the bottom of the post]*
Sharing	*[done by clicking on 'Share' at the bottom of the post]*
Following	*[done by clicking on 'Follow' at the bottom of the questioner's post]*
Showing empathy	Dear *name*, I understand what you mean.
Aligning with the in-group	The most important thing is that we have to spend a lot of time translating all our papers into English; The language assistance services are quite costly and most of us are reluctant to resort to them.
Self-disclosing and sharing personal experiences	Because I am not a native English speaker...; Yes, this has happened to me as well.

[a] Semi-colons indicate a new example.

For instance, one of the threads was initiated by a scholar working at a university in Argentina who did research on the consequences of using English for international publication. This scholar wanted to know the percentage of publications by researchers in countries where English is not the 'mother tongue' that reached their country fellows. The participants contributed in order to provide information on their own publication, often using the metrics provided by RG (i.e. the university of the readers). A month after posing the question, the questioner wrote another post commenting individually on each of the contributions, sharing what he had found (e.g. '*Name*, Thank you so much for answering. Your RG reads for last week shows [...]. Indeed, it is very interesting that 3 of these top-5 countries do not use English as first language'), thanking all the participants (e.g. 'I thank you so much for answering my call') and encouraging other readers to contribute further (e.g. 'More opinions and experiences will be greatly appreciated'). Interestingly, one of the participants, working at a Peruvian university, wrote his post in Spanish, and, when commenting on his contribution, the questioner used parallel language, writing this part of the post both in Spanish ('*Name*, ESPAÑOL: Ojalá más países tuvieran esa determinación'), to express solidarity with this participant, and in English (e.g. 'ENGLISH: I wish more countries had that determination'), to make sure that all the other participants could understand the content.

Due to the disciplinary topics of these threads, 'native speakers' of English frequently take on the role of 'expert' providers of information, and sometimes present credentials for their expertise. For example, when one multilingual scholar asks about the differences between the writing of English native and non-native speakers, in order to help non-native scholars get their papers published, some native speakers comment on 'problems' that non-native speakers have, drawing on their experience (e.g. they proofread papers for scholars who are not native speakers of English, they often co-author papers with non-native scholars). In another thread, when one of the participants who is also an editor states that he sometimes finds 'good science badly analysed or presented', or papers that are 'not well written', another participant asks for clarification, probably to get a better idea of what editors mean with the general comment that the language needs to be improved. The first participant provides a detailed list of the most common problems, thus helping the questioner understand what editors seek (see Example 6.1).

Example 6.1

Participant 1. Dear Profesor *Name*, I am learning from your kind advices, When you said 'good science badly analysed or presented', I highly appreciate if you can share few of your general experience (...)

Participant 2. Name: Very often, the problems are over language (...). The commonest other faults are a muddle among the sections of the paper [*provides a detailed explanation*]. But it is hard to give very general advice.

Participants also write posts to *express their opinion* on issues of academic publishing in English-medium journals that are of particular relevance and importance for them. For instance, they give their opinion on whether multilingual scholars should or need to publish in English or in their local languages. This is the case of the thread with the highest number of posts, 105, which discusses whether there is bias against multilingual scholars' submissions based on geographical location, or whether language issues should be a reason to reject papers for publication. Participants sometimes also *express their feelings, emotions and reactions* to other contributions or towards the issue being discussed (e.g. 'I'm happy to see', 'Your case is surprising and interesting', 'The English issue in the research paper makes me nervous').

Participants also *give advice* or provide suggestions, responding to the questioners' requests. Advice by multilingual scholars is sometimes based on their own experience and on the perception that they conduct research in similar conditions as the questioner (e.g. 'If I receive such comments from the reviewer, I simply request the reviewer for some suggestions to improve', 'Try Zenith Biomedical Editing... they have best price guarantee'). Advice is usually written in a personal way, addressing the advice seeker explicitly (e.g. 'I advise you', 'I suggest you', 'You'll receive useful feedback', 'This can help you').

Many of the functions in these threads have a bonding purpose, that is, they serve to emphasize similarity and common ground between interactants, and to create group identity (see Ädel, 2011). *Agreeing* is used by participants to show that they share the views, perceptions or values of other participants, or to acknowledge that another participant is right. When agreeing, participants sometimes include explicit address of their interlocutor preceded by 'dear' (e.g. 'Yes Dear *name*, that's true'), which indicates respect and social bonding. *Aligning with the group* is a function used by multilingual scholars to highlight shared concerns and create group identification. In posts performing this function, multilingual scholars use the inclusive pronoun 'we', which conveys solidarity and rapport and a sense of belonging, for example, to the 'non-native' group. Participants share their concerns regarding the additional burden, in terms of time and money, of writing their research in English (e.g. 'We have to spend a lot of time translating all our papers into English'), or the difficulty accessing literacy brokers, or 'native speakers' to help them (e.g. 'Many of us from non-English speaking countries may not have these helps easily available'); and their concern because the value of their work may be affected when their research is published in English (e.g. 'We all have so much to say in our native languages. Many times the research loses its value due to translation in English').

Self-disclosing (i.e. making reference to the participants' personal or academic life) also helps scholars to create intimacy and gain support and empathy from other multilingual scholars. In these threads, self-disclosing often involves speaking of one's difficulties as a 'non-native speaker' (e.g. 'Because I am not a native English speaker (…) I experience some difficulty…') or sharing (often negative) experiences when seeking publication. For example, participants use the forum to share their experiences of having their papers rejected because of the language, sometimes after having their manuscript revised by a native speaker, and to express their feelings (e.g. 'I'm clueless'; 'I find it very strange'), which are probably shared by other participants. Self-disclosing often leads others to also share their experiences (e.g. 'This has happened to me as well. I got similar comments (…). And I answered…'), which contributes not only to knowledge sharing but also to reinforcing the sense of in-group solidarity. An interesting form of self-disclosure in these threads is participants' meta-comments about their status as English speakers and about their perceived knowledge of English (e.g. 'My manuscripts are hard to be accepted in reputable journals as my English is poor'; 'My mother language is not English (…) So, I am not so strong in English'). Participants' comments on their 'non-nativeness' helps them to enlist solidarity from other participants, who are mostly non-native English speakers. 'Native English' participants very often acknowledge their status, as a way to justify their attitude (e.g. 'I am particularly sensitive to this: as a native English speaker, I have a kind of privilege'), or to provide credibility to what they are saying, because they frequently present themselves as proofreaders, reviewers or journal editors (e.g. 'I do a lot of polishing of English language for scholars who are not native speakers'). They tend to contribute to the threads in order to provide expert language (and editorial) advice that has been solicited by a non-Anglophone questioner. Language choice is also used by some participants as a self-disclosing technique, mainly to indicate their linguistic background and knowledge of other languages. For instance, in the thread discussing the importance of English as the language of publication, most of the posts were written in English, but a few participants wrote short posts in French or Russian, or wrote the post in English and added a sentence or fragment in a language other than English (e.g. Russian, German) to prove their point and signal their identity (e.g. 'Хотя мы с Вами могли бы поговорить на русском, но кто бы нас здесь понял?', 'Although you and I could speak Russian, who would understand us here?').

Showing empathy is another function intended to create and maintain social bonds between interactants. This is frequently done when participants share negative experiences. For example, after one questioner reported that his submission had been rejected because of language issues, he got supportive responses (e.g. 'You have published so many good research papers in reputed journals. I do not think your language needs English editing services'), which contribute to reinforcing the positive face of the participant by acknowledging his competence as a researcher and pointing to the unfairness of the rejection decision. *Evaluating others' contributions positively* also helps to create rapport with the addressee, by giving credit for a valuable contribution, and with the whole group, by showing interest in the discussion (e.g. 'An interesting topic, and good answers'). *Thanking* and *Responding to thanks* serve to acknowledge appreciation (see Ädel, 2011). *Thanking* is a frequent communicative act because questioners tend to thank those contributing to the thread (e.g. 'Thank you all for your valuable comments'; 'Thanks for sharing your insights on the question'), but contributors also thank the questioner who raised the issue in the forum (e.g. 'Dear *name*, I thank you for asking the question, for this interesting opportunity to discuss'; 'It's good I could find some information (...). Thanks for your question to share, it's really useful!'), or other participants ('Thank you so much for sharing your RG question!').

In addition to writing a comment in response to the question, researchers can also perform acts such as *recommending* or *sharing* questions or responses or *following* the initial question poser. *Recommending* is an evaluative act that not only gives the author of the recommended response credit, but also draws the attention of other participants towards this contribution. By recommending a response, participants increase the number of recommendations displayed (see Figure 6.1) and may contribute to making the response 'popular' and increasing its visibility, as popular responses are displayed at the top of the thread.

As Table 6.1 shows, as a whole, the functions of posts in these threads contribute to building rapport, that is, to promoting concord and social cohesion between the members of the group (see Ädel, 2011: 2933). Although participants expressed their opinion on several issues (e.g. bias against non-native speakers of English; the advantages of publishing in English or the local language), the corpus had no occurrences of explicit criticism and few occurrences of disagreement. In other words, opinions

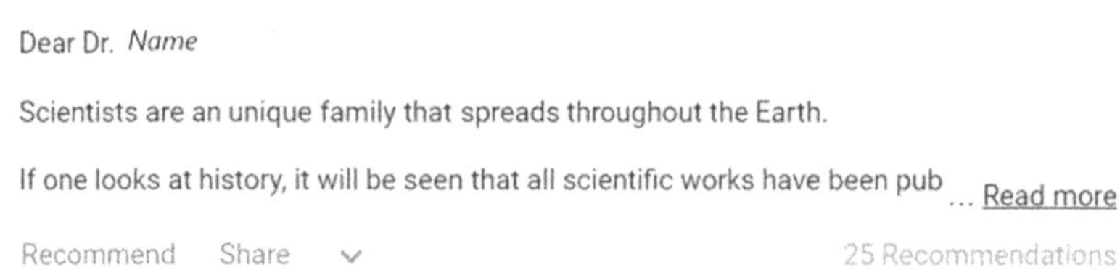

Figure 6.1 Recommending and sharing in RG Q&A posts

were not presented in opposition to those of other participants, but as contributing to the discussion and to sharing experiences related to the issues raised, as solicited by the questioner. This analysis is, however, based on a small data set on a topic (i.e. academic publishing in English) which may trigger more supportive contributions; therefore, it cannot be concluded that participants in RG Q&A threads do not disagree with each other explicitly, or do not engage in face-threatening acts such as criticism. In the data analyzed, we only found one example of explicit confrontation, one that took place between two researchers from the same country (judging from their names and affiliations), when discussing the issue of publishing in English or in national languages. One of the researchers accused the other of distorting and misinterpreting his words and the other answered back, without any mitigation, saying that he had not distorted anything, just quoted the other's words and interpreted them. He then moved on to discuss political issues, rather than language issues. The following contributors to the thread went on to discuss language issues, without making reference to this off-topic post, which lends credence to the participants' interest in engaging in productive, rather than conflictual, discussions.

In sum, the primary function of posting answers to a question in RG Q&A is that of 'sharing' rather than 'replying'. As this analysis illustrates, participants share information, opinions, suggestions, resources or experiences with any interested scholar. They not only answer the questions, but also agree with others, provide support, express empathy and solidarity, recommend responses and justify their opinions drawing on their experience, among other sources of expertise. Our analysis of the Q&A threads in the corpus shows that this digital genre supports new types of interactions and relations among scholars. Previous research on RG Q&A discussions (e.g. Deng *et al.*, 2018) describes them as a valuable source of knowledge for solving research problems. This analysis also shows that the RG Q&A may play a role in the landscape of academic communication as a translocal, cross-disciplinary forum in which to share concerns and opinions, seek and obtain emotional support, and request help and advice in relation not only to research-related issues but also to career advancement issues.

6.4 Concluding Remarks

In this chapter, we have considered the value of ASNSs as multifunctional sites that can be used by researchers for self-promotion, networking, information management and interaction. We have discussed how ASNSs involve new modes of social interaction and new dynamics of knowledge circulation, which is no longer exclusively controlled by scholarly publishers. The sharing of research output through these platforms, however, raises ethical and legal concerns for scholars, derived from the for-profit nature of these sites and the copyright status of some of the publications that scholars self-archive in these platforms. We have also illustrated the role of RG Q&A forums as globalized online affinity spaces where ELF is used by (multilingual) scholars not only to exchange information but also to request and provide help and advice and share their concerns and experiences regarding academic issues with other researchers across the globe. These forums therefore fill the niche for an open and fluid space where scholars with different degrees of expertise can tap collective intelligence informally and also turn to for support and help, reflecting a socially bonded multilingual academic community.

Notwithstanding the potential of ASNSs for enhancing collaboration and sharing, knowledge production and circulation, there is a need for more research on whether these sites actually facilitate a more equitable access to knowledge; whether the commercial nature of these sites and the commodification of the data generated in them have an effect on their use by researchers; whether researchers' engagement with this type of peer-to-peer sharing is leading to changes in scholarly publishers' practices of open access and is actually transforming the traditional model of scientific publishing. There is also a need to further investigate multilingual scholars' language practices in Q&A forums and their experiences and perspectives regarding participation in these forums. At present, little is known about the role of multilingualism and about ELF and language(s) choice in researchers' engagement with this type of peer-to-peer sharing. This is worth investigating given that ELF users are the predominant sociolinguistic group in today's research world.

In the next chapter, we discuss genres used to communicate research informally to a diversified audience, where the erosion of boundaries between local and global audiences leads to multilingual practices.

Notes

(1) For the question and responses in each thread, we retrieved each participant's name, institution and the content of the post. To protect the interactants' anonymity, no other information from participants' profiles was collected or analyzed.

(2) Although 'native' is a contested term in relation to linguistic affiliation, we use it in this chapter because it is the term used by participants in the Q&A threads.

7 Disseminating Knowledge to Diversified Audiences

7.1 Introduction

Public communication of science has become a key activity for scientists and the new online media offer suitable platforms for public engagement and for interaction and dialogue between scientists and society. In social media platforms contexts usually collapse, that is, multiple audiences are flattened into one (boyd, 2002; Marwick & boyd, 2011), and the line between communication with peers and communication with other publics becomes blurred (Trench, 2008). Although collapsed contexts enable researchers to reach multiple audiences simultaneously, they also challenge writers to manage communication across diverse sets of norms and expectations (boyd, 2002) and to employ strategies for making knowledge understandable and cater for the communication needs of these diversified audiences.

There is increasing research on the possibilities offered by digital genres and social media for reaching a wide audience and promoting scientific literacy in the lay public and on the rhetorical work of these genres to accommodate various audiences. For instance, López-Pérez and Olvera-Lobo (2016) study how research institutions use Facebook, Twitter and YouTube to communicate scientific results to society, and Hara *et al.* (2019) examine interactions between lay people and scientists through Ask Me Anything (AMA) sessions on Reddit's 'Science' subreddit (r/science). In these genres, knowledge that was originally produced in a specialized context is recontextualized to make it comprehensible and relevant for a wider public (see Calsamiglia & Van Dijk [2004] for a detailed account of recontextualization).

Given their high popularity among researchers and organizations as genres for communication with diversified audiences, in this chapter we focus on blogs and microblogs (Twitter). We first briefly review research on the use of blogging and microblogging to disseminate knowledge and on the strategies used by bloggers. We then present a case study on the Twitter accounts of six research groups to illustrate the interplay among genres, audiences and languages.

7.2 Science/Academic Blogging and Microblogging to Reach Diverse Audiences

The emergence of science/academic blogging may be accounted for by new perspectives on scientific research and communication brought about by the paradigm of open science (see Chapter 3). Trench (2008: 185) argues that online genres are blurring the boundaries between scientific and public communication and 'turning science communication inside-out', pointing out that due to their personal character, blogs facilitate this 'inside-out' process. This potential of blogging for reaching broader audiences has been harnessed not only by individual researchers but also by academic journals, such as the *British Medical Journal* (*BMJ*) or the *Public Library of Science* (*PLoS*) journals, and many science organizations and institutions, for which knowledge dissemination is of pivotal importance (e.g. the charity Cancer Research UK, the University of Oxford). Twitter has also become an important tool for research discussions and disseminating knowledge to the public, also embraced by journals and institutions. The *BMJ* and *PLoS* journals, for example, have several issue-specific Twitter accounts (Bjerglund & Söderqvist, 2012). An increasing number of scientific societies, academic institutions and research centers are using Twitter to promote their research and activities, sharing resources and disseminating information to the public (Lópéz-Goñi & Sánchez-Angulo, 2018).

As platforms for communicating with a broader public, blogs and microblogs share many functional features. Both allow users to access a variety of content including private and public content, news, opinions and reflections; they both support interactive features, which facilitate dialogue and discussion (Smith, 2015); and both may contain links to other web pages, which may motivate readers to click and access longer documents with more detailed information. However, there are also differences, which may lead to different uses (Kaplan & Haenlein, 2011; Smith, 2015). Microblogs facilitate synchronous connections and real-time conversations made up of short exchanges among individuals and groups, while blogs facilitate conversations and discussions with longer and more detailed contributions, which can be traced back through comments. As Twitter enables writers to tell others what they are doing at a particular moment, it creates 'ambient awareness' (Zappavigna, 2012) and can trigger feelings of closeness and intimacy (Kaplan & Haenlein, 2011). Users can 'follow' relevant or interesting Twitter accounts, which reduces the effort of accessing information, a particularly important feature for reaching the lay public (Smith, 2015).

7.2.1 Blogs

Two overlapping terms referring to blogs that discuss academic and science-related issues are 'academic blogs' and 'science blogs'. The basic criteria to define 'academic blogs' are their content (i.e. academic- and discipline-related topics) and authority (i.e. written by a person with expertise in an academic field) (Luzón, 2011, 2017). 'Science blogs' are defined only in terms of content as 'blogs for science communication' (Gardiner *et al.*, 2018; Mahrt & Puschmann, 2014; Ranger & Bultitude, 2016). They may be written by scientists, teachers, university students, postdocs or science journalists, although they are frequently written by active scientists (Gardiner *et al.*, 2018; Jarreau, 2015).

A large number of studies have investigated the audiences of science and academic blogging and the discursive strategies employed by bloggers. Most researchers characterize blogs as platforms to share knowledge and interact with multiple audiences, that is, peers, amateur scientists, students, practitioners, interested public (e.g. Jarreau, 2015; Wilkins, 2008). Blogs can be used as (i) a mode of informal communication among experts, not constrained by the conventions of other academic genres (Heap & Minocha, 2012; Kjellberg, 2010); and (ii) as a way to improve outreach and communicate academic and research related information to non-experts (Buehl, 2016; Luzón, 2013).

Some studies have highlighted the role of blogs as a new genre to communicate with peers. Saunders *et al.* (2017) distinguish between 'science communication blogs' (i.e. blogs used to disseminate scientific information to the general public) and 'science community blogs' (i.e. blogs aimed at other scientists, focusing on issues related to academia and the process of science), although they acknowledge that there is an overlap between them, 'with some blogs addressing both audiences with a mix of post topics and writing styles' (Saunders *et al.*, 2017: 3). By using examples from their own science community blogs, they analyze the value of these blogs for the community of working scientists. They argue that, among other benefits, these blogs help the rapid communication of ideas, observational notes and advances in teaching and research methods; and provide a forum for the discussion of personal experiences which can help junior researchers and scientists from underrepresented groups. Mewburn and Thomson (2013) suggest that academic blogs can be regarded as 'virtual staff rooms', where academics can engage in conversations involving personal and professional issues. Kouper (2010: 3) also compares blogs to 'a virtual water cooler for graduate students, postdoctoral associates, faculty, and researchers': blogs provide these users with a platform where they can share news and information on academic topics.

Other studies emphasize that blogs may help to fill a gap in knowledge sharing, since they are used to communicate with multiple audiences and reach publics beyond academia (Luzón, 2013; Mahrt & Puschmann, 2014; Mehlenbacher, 2019b). Heap and Minocha (2012) argue that one of the main benefits of blogs for scholars is that they serve to disseminate information and research openly and without the constraints of formal publishing, which make them accessible and comprehensible not only to peers but also to the interested public. Thorsen (2013) explains how researchers working in Antarctica used their blogs to act as citizen journalists and report on the effects of global warming first-hand, without the mediation of mainstream media, to show how the scientific community is dealing with this problem and to bridge the gap between science and lay people. Bloggers (both researchers and science writers/journalists) in Jarreau's (2015) study also perceived themselves mostly as science 'explainers' (i.e. making science understandable to non-specialists), public intellectuals (i.e. drawing attention to topics relevant to society) and civic educators.

Research on science/academic blogs has also revealed the discursive strategies employed by bloggers to cater for complex audiences and engage with the interested public (Luzón, 2013; Mehlenbacher, 2019b; Sokół, 2018). Our study on posts used to comment on published research related to a disciplinary area (Luzón, 2013) analyzed the discursive strategies used by bloggers to recontextualize scientific discourse in the realm of science blogs. The study revealed that bloggers use a variety of strategies to (i) frame content for both expert and non-specialist audiences (e.g. explanation of terms and concepts, visuals, links to sites providing clarification of concepts for non-experts or additional specialized information for peers); (ii) engage readers by arousing their interest and using interaction to create intimacy and immediacy (e.g. self-disclosure, conversational discourse, inclusive pronouns, reader references, humor, questions); and (iii) by evaluating research and findings with the purpose of influencing readers. These strategies suggest that bloggers perform the role not only of experts who share information with peers and with the public, but also of civic scientists who make scientific research understandable to the interested public or discuss scientific research to influence societal issues (Kyvik, 2005). Zou and Hyland (2019) analyzed blogs of academics in the social sciences to determine how they used stance and engagement to recontextualize the information they had recently published in journal articles. Like Luzón (2013), they found that when writing blogs academics adopt a more informal and engaging style so as to reach a wider audience. In her study of posts from 16 blogs from the *PLoS* blogging network, Mehlenbacher (2019b) found that the rhetorical strategies used by bloggers are borrowed both from genres used to communicate with experts and from popular science texts, which confirms previous findings on the diversified audience and the hybrid nature of scholarly and science blogs.

Researchers have also explored how comment threads facilitate discussion and public engagement (Bondi, 2018; McGrath, 2016; Pigg *et al.*, 2016). Pigg *et al.* (2016) analyzed the interaction in the comments of the post 'The Chicken and the Egg' in *Science Buzz* (a community blog facilitated by the Science Museum of Minnesota). The analysis of the discursive moves in this online conversation, where participants ask and answer practical questions, revealed that online social media can be spaces for knowledge exchange and public science engagement. Shanahan (2011) also examined two posts in a single blog, *Not Exactly Rocket Science*, and shows that they act as forums for interaction and collaboration between a scientist and a farmer. McGrath (2016) analyzed comments to five posts in the blog *Polymath*, through which participants, both experts and non-experts (in this case, students of mathematics, or scholars from other disciplines, rather than the general public) collaborate in the writing of a research paper. Bondi (2018) shows that, thanks to the possibility for readers to engage in open commentary and discussion, blogs are sites where knowledge can be both disseminated and co-constructed. Her analysis of a small-scale corpus of blog threads in posts written by well-known economists reveals that bloggers establish dialogicity by using engagement devices and inviting the readers to react in the post. The open participatory framework of blogs makes it possible for readers with different degrees of expertise to participate by contributing to one of the parallel conversations in the ongoing debate that interests them, which creates the impression of interwoven polylogues.

Despite the potential of blogs for informal communication and dialogue with a heterogeneous audience, they also pose challenges to researchers and impose constraints on the discussion of research. There is no clear or single institutional response to academic blogging and, despite the importance that funding bodies attach to the impact of research on society, blogging is not usually recognized as a form of scholarship and, unlike high-stakes genres such as journal articles and books, blogging to date has little value in research assessment and in promotion and tenure processes (Heap & Minocha, 2012; Mewburn & Thomson, 2013). Blogs that are open to the public also raise concerns about the risk of revealing confidential information or information that could be used by others without the blogger's permission (Barton & McCulloch, 2018; Heap & Minocha, 2012). Other important issues to the researchers producing blogs are the time required to write them, vulnerability (i.e. mistakes may be more visible) and the difficulty of removing content later on (Heap & Minocha, 2012). Fahnestock (2016) also problematizes the role of blogs as tools for the public's participation in scientific discussions. She points out that the interactivity enabled by the comment/response feature of blogs may result not only in substantive responses, but also in tangential comments and uninformed opinions by non-experts, which may prevent the creation of an informed public.

7.2.2 Twitter

Twitter has become popular among scholars as a tool for self-promotion, information sharing and networking, generating an increasing interest in the use of Twitter to disseminate scholarly and scientific information. Scholars have emphasized the potential of Twitter as a tool used by researchers to speed up the pace at which they communicate within their disciplinary community and to disseminate knowledge to the general public (Bik & Goldstein, 2013; Darling *et al.*, 2013; Mehlenbacher, 2019b). In their study of how scientists from a variety of academic disciplines use social media, including Twitter, Collins *et al.* (2016) found the following benefits (as reported by participants): the size and diversity of the potential audience; the possibility to communicate 'snippets'; the short amount of time needed; the accessibility of the content; and its potential for networking with other scientists and for communicating directly with the public.

Focusing on their discipline of marine ecology and conservation, Darling *et al.* (2013) discuss the usefulness of Twitter for communicating with different audiences at the various stages of a scientific publication. Twitter provides a platform through which colleagues around the world can help to generate and refine new ideas; it can also be used 'as an informal arena for the pre-review of works in progress' (Darling *et al.*, 2013: 32). After publication, research results can be tweeted to reach a wide audience, including peers, decision makers, the media and the interested public. 'Citation tweets', that is, tweets that contain a link to a publication, contribute to increasing the outreach of the publication, as Twitter provides an 'echo chamber' (Darling *et al.*, 2013: 37) for its dissemination, thus potentially enhancing its scientific and social impact. Twitter is also often used in conferences for 'live-tweeting', that is, tweeting key points and information from presentations in real time (Shiffman, 2012). Live-tweeting contributes to increasing the outreach of the contents of the presentations (Sugimoto *et al.*, 2017) and allows attendees and non-attendees to discuss issues related to the presentations and to build networks (Darling *et al.*, 2013). In addition to tweeting to inform about new publications or to interact in conferences, researchers use tweeting to interact with peers for other purposes, for example to share resources and information related to their professional practice and their teaching, request assistance from and offer advice to others, engage in social interaction, network with others and construct an online identity (Veletsianos, 2012).

Even if Twitter may be most frequently used for communicating with peers, its affordances and technical features also make it suitable for reaching the public and making research results available beyond the disciplinary community (Kahle *et al.*, 2016; Shiffman, 2012; Smith, 2015). Twitter is a particularly useful tool for raising awareness of important social issues and for the timely and fast communication of scientific issues that require urgent social action. Research has shown that, in communication with the general public, tweets may serve two broad communicative functions: information dissemination and education, and calls for action. Jahng and Lee (2018), for instance, report on the use of Twitter by a team of scientists (Flint Water Study team) to inform the public about the contamination of water with lead in Flint and call for collective action. Park *et al.* (2013) also found that major health organizations (e.g. the American Heart Association) used Twitter to disseminate health-related information and to encourage actions to promote personal health. The use of Twitter for outreach is particularly important in some disciplines (e.g. conservation science, climate science, health-related disciplines) (Bombaci *et al.*, 2016; Shiffman, 2012), where public support is essential for changing government policies or in fields where the interested public participates actively in knowledge production (e.g. paleontology) (Bex *et al.*, 2019).

However, the potential of Twitter to reach the interested public does not seem to have been fully realized (Smith, 2015). For this reason, several scholars have examined the impact of Twitter as a tool for science communication beyond academia and explored how Twitter can be used for this purpose. Didegah *et al.* (2018) analyzed the quality of the interactions and public engagement related to scientific papers on Twitter, and found that most tweets just disseminated the papers by including the paper title or URL, but tweeters rarely added extra information or comment on the paper in order to generate dialogue and engagement. In a study of live-tweeting at the 2013 International Congress for Conservation Biology (ICCB), Bombaci *et al.* (2016) found discrepancies between the intended audiences and those that were actually reached through live-tweeting. For instance, although a relatively high number of presenters wanted to reach government agencies, policymakers and non-governmental organizations (NGOs), these audiences rarely read these tweets. The authors suggest that providing Twitter-friendly summaries of the presentation with relevant hashtags and usernames could help to communicate the content accurately and broadly. In their analysis of CERN's media platforms, Kahle *et al.* (2016) found that some techniques were particularly useful for generating public engagement. These include images and animation, embedded videos or virtual tours, comments and discussion, 'clickbait' (i.e. strange images or enigmatic texts intended to attract the reader's attention and encourage them to click on a link), tailored content (links leading to content in the reader's local language) or human stories. These studies reveal that although Twitter is a promising tool for scientists to engage directly with the interested public, researchers still need to be made aware of appropriate strategies.

One aspect to which little attention has been devoted, with a few exceptions (Luzón & Albero, 2020; Orpin, 2019; Walter *et al.*, 2019), is the semiotic devices (including language) that researchers use to make meaning in their tweets and reach various publics. In their analysis of scientists' tweets on climate change, Walter *et al.* (2019) found that they use language strategically, adjusting it to different audiences. While their language is quite neutral when communicating with peers, there is a higher frequency of words expressing negative emotions when communicating beyond the scientific community (e.g. with civil society actors or journalists), probably in order to stress negative consequences. In their study of tweets posted by participants in two conferences, Luzón and Albero (2020) also examined how different semiotic resources (e.g. linguistic forms, pictures, videos, embedded slides) are combined to make meaning in informal communication with peers. They found that in order to achieve the rhetorical purposes of these tweets, scholars compose them by using a variety of (linguistic and non-linguistic) expressions of stance and engagement (Hyland, 2005).

7.2.3 Choice of language in blogs and microblogs

Another under-researched issue is the language in which knowledge is communicated through blogs and tweets, and the linguistic and cultural background of bloggers and tweeters. Most research on academic blogging has focused on blogs written in English and mainly from English-speaking Western countries. Only a few studies have sought to analyze blogs written by non-Anglophone scholars in their local languages or in English (e.g. Kjellberg, 2009; Luzón, 2018b, 2019a; Sokół, 2018). Sokół (2018) studied blogs run in Polish by individual healthcare practitioners; Kjellberg (2009) studied Swedish academic blogs (including blogs in Swedish and a few blogs in English); and Luzón (2018b, 2019a) analyzed blogs written by multilingual scholars who wrote their blogs exclusively in English as a lingua franca or combined English with their local languages. The scarcity of studies on academic blogging by non-Anglophone scholars may be due to the linguistic background and knowledge of the analysts, as suggested by Mewburn and Thomson (2013), who regretted that they lacked the skills to analyze blogs in languages other than English. However, the reason could also be that these blogs seem to be fewer in number and less popular. For instance, for their study of blogs in ResearchBlogging.org (and aggregator of research posts from blogs in several languages), Shema *et al.* (2012) chose blogs with a minimum of 20 entries within the span of a year, to make sure that they were fairly established, but only 14% of the blogs that satisfied these criteria were written in languages other than English.

There is abundant research on the variables that influence the language that scholars in non-Anglophone countries choose when writing traditional genres, for example, the target audience, or the genres used for dissemination in different disciplinary communities (Duszak & Lewkowicz, 2008; McGrath, 2014). Research on the publishing and writing practices of multilingual scholars has also evinced that one of the challenges faced by many of these scholars is the need to combine publication in English with local knowledge production (Curry & Lillis,

2018). Given the possibilities afforded by blogs (and other digital genres) for multilingual communication of research, they may be sites in which to negotiate the tensions between the need to achieve international visibility (without the constraints of gatekeeping) and the desire to reach a broader local audience.

However, the issue of language choice when writing scholarly blogs, the consequences of this choice and the challenges researchers might face has received little attention. Even in studies that explore the blogging practices of scholars from non-Anglophone linguistic backgrounds, language choice is either not mentioned or only discussed briefly. In Kjellberg (2010), Scandinavian bloggers stated that the choice of blogging language was influenced not only by the target audience, but also by the desire to gain practice in writing in English, or to maintain fluency when writing in their local languages. One of the respondents accounted for her choice of Swedish by stating that as she was using English for her academic activities, she felt she was losing contact with her local language. Sajeev *et al.* (2019) explore the blog *Climate Footnotes*, which is written mainly by authors based in Austria and Germany, and mostly in English, with a few posts in German. The authors point out that a motivation to write the posts in languages other than English is to make it easier for the local community to understand scientific research which has a strong local focus. They illustrate this point with the post 'The pressure is on! – About the European cold spell in spring 2016'/'Nur nicht zu viel Druck! – Der Kälteeinbruch im Frühling 2016', which was simultaneously published in English and German, with the German version having almost 50% more views. Thorsen (2013) analyzed more than 50 blogs written by scientists working in Antarctica during the International Polar Year (2007–2008) to communicate directly with the general public and report on the effects of global warming. Most of these bloggers wrote in English, which reveals that they were aware of their (potential) global audience. However, there were also many blogs in other languages (e.g. French, German, Norwegian and Spanish). Thorsen (2013) exemplifies the relationship of language choice and intended audience with the case of a German blogger who started the blog in English but switched to German after his German friends sent 'severe protests' via email. One comment on his second entry also urged the blogger to use German: 'BITTE SCHREIB DEUTSCH! ...vergiss nicht deine wurzeln im eis!' ['PLEASE WRITE GERMAN! Do not forget your roots on the ice!'].

Luzón's (2019a) study of academic blogs by multilingual scholars focused on language choice and revealed that English and other languages are combined in different ways in these blogs, with the most common pattern being a mixture of posts in English and posts in the blogger's local language in the same blog. English was used to reach a wider global audience while the local language was used to address readers who shared this language. The alternation of languages in these blogs enabled bloggers to participate in different communities, to adjust their code(s) to heterogeneous audience(s) and to mobilize their linguistic resources for different purposes. In blogs written by research groups affiliated to Spanish institutions (Luzón, 2018b), English was also sometimes combined with the scholars' local languages (Spanish and/or Catalan). An analysis of the posts written in English revealed that many contained (to varying extents) examples of non-conventional forms of the language, which suggests that lexico-grammatical variation is more accepted than in non-digital academic writing, where there is usually little tolerance of non-standard forms (Hynninen & Kuteeva, 2017; Lillis & Curry, 2010). Although research on academic writing reflects a trend towards an increasing acceptance of non-standard linguistic forms in some high-prestige international journals (Rozycki & Johnson, 2013; Tribble, 2017), linguistic correctness when writing in English for publication purposes is highly valued, even by multilingual scholars (Hynninen & Kuteeva, 2017), and the language of manuscripts is often revised and negotiated in the reviewing process (Flowerdew & Wang, 2016). In the absence of this gatekeeping control in academic blogging, scholars are less constrained by Standard English norms, and they use 'non-canonical' forms and draw on multilingual resources to meet their communicative needs.

The choice of English to reach an international audience suggests that English is the lingua franca for informal knowledge communication on the internet. This claim is supported by Yu *et al.*'s (2018) study of the language distribution of scientific tweets and Kahle *et al.*'s (2016) study of CERN's social media platforms. Yu *et al.* (2018) found that English was the dominant language in tweets about publications indexed by Scopus. Kahle *et al.* (2016) revealed a much larger audience for CERN's Twitter account in English than for its French account (1.03M vs. 12.2K). Barata *et al.* (2018) also point to the dominant role of English in distributing scientific information globally on the internet. After the Zika virus outbreak, the authors examined the language in Twitter and Facebook posts that linked to Zika-related research to determine the audiences reached by these tweets. Most of these posts on Twitter (up to 90%) and Facebook (76%) were in English, even though Brazil was the most affected country. However, local languages are also used in the scholars' tweets, especially to communicate science and promote scientific literacy among the local audience outside academia. Spanish, for instance, was used by a group of members of the Spanish Society for Microbiology (SEM) to teach a course of General Microbiology, emitted with the hashtags #microMOOCSEM and #microMOOCSEM2. Interestingly, 58% of the followers of the second course were located outside Spain (López-Goñi & Sánchez-Angulo, 2018), which suggests that Spanish also has a role as a language for international communication online. At the time of writing, the SEM is using its Twitter account to disseminate scientific information about Covid-19. Although the tweets are in Spanish, they embed and link to documents in Spanish and English, revealing the multilingual nature of tweeting by researchers from non-Anglophone linguistic backgrounds.

Despite evidence that multilingual scholars use both English and other languages in their tweets, little attention has been paid to how these scholars use their Twitter accounts to reach diverse local and global audiences (peer researchers, students, interested publics) and how they strategically choose languages for this purpose (i.e. choosing between English, their local language or other languages in their linguistic repertoire).

7.3 Research Case Study: Networked Language Practices of Research Groups on Twitter

In this section, we present the results of a study in which we analyze how research groups harness their Twitter accounts to keep different audiences informed of their research activity and results. We describe the networked language practices of research groups affiliated to Spanish institutions when using Twitter. We analyzed how these groups employ their linguistic repertoires and engage in multimodal digitally mediated practices to connect with diverse audiences and negotiate their relationships with these audiences. We address these questions: (i) For what purposes do research groups use Twitter?; (ii) What are the language choices of Spanish research groups in their Twitter accounts in relation to their networked audiences?; and (iii) How are different semiotic resources combined to connect with different audiences and achieve the groups' purposes?

We combined content and discourse analysis of 600 tweets taken from the Twitter accounts of six Spanish research groups in the fields of engineering, medicine and chemistry. To be selected for the study, the research groups' Twitter accounts had to meet two criteria: (i) active accounts with regular tweets and a relatively high number of followers (more than 150, although some of them had many more); and (ii) all or some of the tweets should be written in English. To compile the corpus, we took 100 tweets from each account working backwards from March 21, 2019. These Twitter accounts are multilingual sites, both because in most cases the groups tweet in both English and local languages (Spanish, Catalan or both), and because they retweet and embed documents in several languages (mostly English, Spanish or Catalan). Each tweet post was coded for its main communicative purpose (i.e. function) and the language of the tweet. The tweets were then examined to analyze the interplay of different semiotic resources (linguistic forms, images, audiovisuals, gifs) or Twitter features, such as hashtags or mentions, used to make meaning and achieve several rhetorical purposes. For the analysis of linguistic forms, we focus mainly on engagement and stance features (Hyland, 2005). Both language choice and the functions of the tweets vary considerably across accounts and depend on the communication agenda of the research groups. For instance, while some groups write all their tweets in English, as a way to index their international identity, others combine tweets in different languages.

The tweets have various communicative functions (see Table 7.1), which contribute to more general purposes: (i) community building and networking; (ii) self-promotion and publicizing of research output; (iii) calls to action; and (iv) public outreach and dissemination. Tweets posted to fulfill these rhetorical functions may be intended for different audiences, which influences the language used to compose them.

Many tweets are used by the groups to perform actions that contribute to establishing links with the academic community by discussing shared concerns, collaborating and sharing resources or providing evidence of their links within this community. The groups commented on aspects of academic life, such as publishing, reviewing or open access, and used language and other semiotic resources to express solidarity. In Figure 7.1, for example, the inclusive 'we' in 'We love to pay double' contributes to the idea that this complaint is shared by the members of the academic community; the emoji emphasizes the feeling of frustration and irritation. The groups also engaged in disciplinary or academic humor, for example, by embedding funny gifs, discipline-related images/jokes or discipline-related versions of popular or viral videos. Often, they retweeted humorous messages, which helps to create links within the community.

Another community building use of tweets is the sharing, through linking, of peer-reviewed publications and resources, as well as practical

Table 7.1 Purpose of tweets published by Spanish research groups

(i)	**Community building and networking**
	Commenting on aspects of academic life
	Sharing information, news and concerns about professional careers
	Engaging in disciplinary or academic humor
	Sharing resources, publications and disciplinary information
	Helping other groups to spread information by retweeting
	Engaging in positive public evaluation of others
	Making their disciplinary links public
(ii)	**Self-promotion and publicizing of research output**
	Notifying the publication of a new paper
	Informing of their members achievements
	Sharing their own resources and research deliverables
	Announcing their talks in a conference
	Commenting on their international projects
	Informing of their presence in the media and providing access to the document
	Informing about their dissemination activities
	Commenting on their research and findings
(iii)	**Calls to action**
	Encouraging students or scholars to register or take part in academic activities
	Encouraging the general public to take civic action
	Encouraging authorities and policymakers to take specific measures
(iv)	**Public outreach and dissemination**

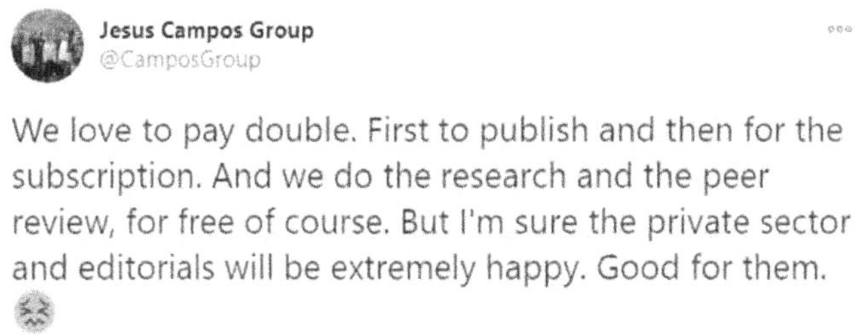

Figure 7.1 Tweet commenting on aspects of academic life

information that may be of interest to other members of the community. The groups also indicated their willingness to collaborate by retweeting information that other scholars would like to spread and share. Retweets have been considered as mechanisms for quotation, or 'internal citation' (Weller *et al.*, 2011: 4), which facilitates the diffusion of information (Puschmann, 2015b; Zappavigna, 2012). The research groups retweeted tweets which advertised positions in other institutions or groups, announced study grants, informed about other scholars' publications or dissemination activities, or informed about events organized by other groups, among others. In Figure 7.2, several users (among them the group that is retweeting) had been sent the information through the @mention feature and had been requested to retweet and spread, with the use of the imperative form 'propaga' (*spread*) and the vernacular form 'Po favó' (*please*), which produces a humorous effect and contributes to reinforcing solidarity bonds.

The research groups also used Twitter to engage in positive public evaluation of others and to acknowledge others' positive evaluation of their own work. They composed tweets where they praised other researchers' work by using evaluative vocabulary ('great talk', 'amazing talk') and included a picture of the event in question, or where they congratulated others on an achievement and retweeted posts that provided more information on the achievement, which contributes to enhancing the prestige of the 'congratulated' scholar.

Groups also tweeted to make their disciplinary links public. Members of the groups informed about their participation in social events with others (either by tweeting about them themselves or retweeting), about academic invitations (e.g. to participate in a conference, to visit a research group) or about their collaboration with other scholars (e.g. visits by other scholars). These tweets often included photographs as evidence of what is said in the text.

Regarding language choice, although most of the networking tweets were written in English, some were written in local languages, for example, when the purpose was to maintain links with members of the

Figure 7.2 Tweet to help other groups spread information

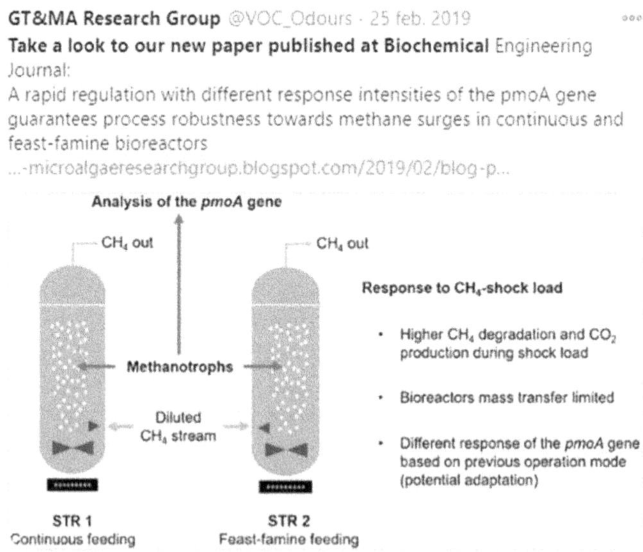

Figure 7.3 Tweet notifying that a new paper has been published

local disciplinary community, or discuss aspects of local academia (e.g. complaining about the situation of researchers in Spain).

Self-promotion is another important objective of the groups when writing tweets. The primary purpose of many of these tweets is to advertise the group's work and achieve visibility. Several communicative functions serve this purpose (see Table 7.1). Research groups used Twitter to advertise their own research, events, publications and achievements, and to encourage people to access their publications and online resources. When informing about a new publication, they might include only the link, but they sometimes include a short text or graphical abstract. These tweets tend to include an imperative, informally encouraging peers to read the paper (e.g. 'take a look to') (see Figure 7.3).

The tweets informing about the group members' achievements (e.g. PhD defenses) and participation in academic events often included a positive evaluation of the member's performance (e.g. 'great presentation and discussion') and pictures of the event. They also tended to include congratulations (e.g. 'Shout out for Dr Shen!!! ICIQ's newest #doctor has successfully defended his #thesis today. #Congratulations!') and emojis indicating celebration. Given the increased awareness of the importance of outreach, research groups also promote themselves by informing about their dissemination activities and showing the relevance of their research for society.

The tweets in the corpus announcing new publications were written in English, probably because they were intended for the international disciplinary community. As for the other promotional functions, language choice depends on the group's agenda and the stakeholders for whom the results might be interesting. For some of the groups, the audience seems to be not only peers, but also practitioners and even interested readers. This is the case, for instance, of *SOLTI*, a group doing research on breast cancer, which uses Twitter to communicate with practitioners, keeping them informed of their research and training courses. When the intended audience is local (the interested public, local practitioners), the group uses their local language, as illustrated in Figure 7.4.

Tweets that informed about the groups' dissemination activities or presence in the media (intended for the local public, including local research funding bodies) were usually written in the local language(s). The local language was also chosen for tweets where groups comment on the relevance of their research for the local economy, as is the case of many tweets written by the *Irrigation Engineering and Management Research Group (AGROENG)*.

The purpose of many tweets in the corpus is an explicit or implicit *call to action*. Readers are encouraged to take different forms of material action: register for a degree or course, apply for a PhD position, attend a conference or follow the group in other social networks. Many of these tweets include links so that readers can click on them and perform the action, imperatives which direct the reader to carry out specific actions (e.g. 'apúntate', 'Contact me') and questions

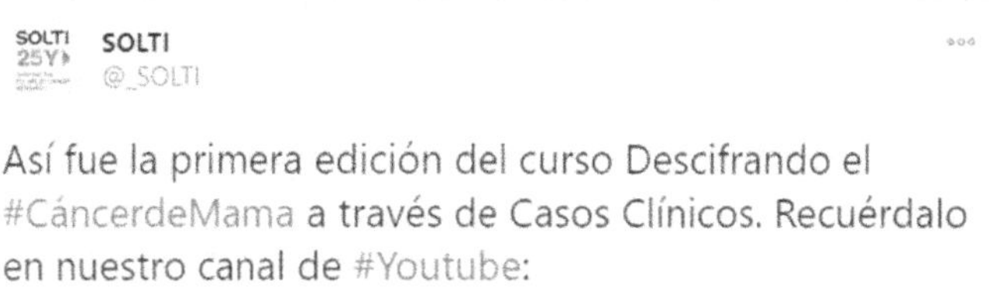

Figure 7.4 Tweet used to inform about the group's dissemination activities

which suggest a situation that would be desirable for the readers ('Quieres unirte a nuestro objetivo?'). Some groups also use Twitter to ask the general public to take civic action, usually by retweeting tweets from NGOs or other users, or to ask policymakers and authorities to take some measures. The language of tweets calling to action depends again on the intended audience. For instance, tweets advertising undergraduate and master degrees and urging students to register are written in Spanish or Catalan, while tweets advertising PhD positions and encouraging students to apply for them are written in English so as to attract international students. Tweets intended to reach local authorities or urging practitioners to do something (e.g. attend a course) are written in Spanish, while tweets calling for conference papers are written in English.

Finally, some tweets are used for public dissemination and outreach, to share scientific knowledge (not only about the group's own discipline) with other scholars and with the interested public. The groups commented on newly published findings which may be of interest to the general public in a non-technical language (see Figure 7.5), or included popular science texts (e.g. a document explaining the different types of ecological food), by embedding them, linking to them or retweeting. Since the main target audience is the interested public, the local language is used, although groups may retweet or link to documents in English. Interestingly, it seems that, despite the new affordances of Twitter for multinational and multilingual communications, the researchers in this study tend to reproduce assumptions of linguistic orders (with English for the international audience and the local language for the local audience) that are in play in other forms of research publication (e.g. Lillis & Curry, 2010).

An important factor influencing the choice of language in these tweets is the academic discipline. Groups doing research in chemistry (mostly basic rather than applied research) tend to write most of their tweets in English, suggesting that they do not specifically target the local audience. By contrast, the group *AGROENG*, which does research on water

Figure 7.5 Tweet for public outreach and dissemination

resources in agriculture focusing on the local context, writes most of its tweets in Spanish or Catalan, indicating that its main audience is local.

This study reveals that the coexistence of languages in the Twitter accounts of these groups results from the fact that they combine languages as a tactic to target different audiences, and that they retweet tweets (and embed documents) in the different languages in their linguistic repertoire, assuming that they will be comprehensible and understandable by at least some of their followers. The case study also shows the situated nature of research group tweets. The choice of language in these tweets may be determined by several related factors (audience, discipline, topic and function of the tweets), but this choice is individualized, depending on different factors for the different groups, and even on the member of the group writing the tweet and his/her orientation to a particular audience at the moment of writing. Given that language choice seems to be highly dependent on the particular group's agenda, complementing the analysis of the tweets with interview data would have admittedly offered deeper insights into these groups' perceptions and attitudes toward the academic languages in their linguistic repertoire, and into the motivations underlying their language choices.

7.4 Concluding Remarks

In this chapter, we have illustrated the important role of academic blogs and microblogs in opening up science for a diversified audience, making scientific knowledge available and conceptually accessible to various stakeholders, including the interested public, and enabling two-way interaction between researchers and their audiences. The chapter has also discussed the discursive strategies used by the writers of these genres when they seek to reach simultaneously specialized audiences and other publics, and the languaging practices they engage in, when contexts collapse, in order to reach these heterogeneous audiences. Here, as seen in this case study, multilingualism plays an important role. Academic blogs and tweets are new genres that respond to new social needs: to communicate informally (and more rapidly and widely than through formal publications) with peers, which fosters collaboration, and to broaden the impact of research by making it available to various interested audiences.

The chapter has also discussed, even if briefly, the challenges posed by these genres for researchers and for scientific progress, e.g. the lack of recognition as forms of scholarship, the vulnerability of some bloggers, the persistence of incorrect content and the validity of the information. We have also seen that although Twitter can be effectively used to disseminate knowledge, researchers do not always manage to engage a wide audience and may need to improve their digital literacy skills to harness the affordances of this genre for communication with the interested public.

The genres discussed in this chapter may assist scholars in achieving the goals of the public school of open science by making research results available and understandable for the interested public. Chapter 8 also focuses on genres used to open up research, but the genres discussed are intended for more active engagement of the public.

8 Engaging the Public in Research

8.1 Introduction

At the onset of this volume, we underlined the impact of the technological developments of the internet on the production and dissemination of academic knowledge. In subsequent chapters, we have also illustrated and discussed different genres with which researchers today are engaging in new discourse practices – ranging from open laboratory notebooks to parascientific genres such as research blogs and microblogging. This chapter explores other emerging discourse practices that also rely on digital affordances but that, unlike the previous ones, involve engaging audiences in research processes. While blogs and microblogging enable researchers to communicate and exchange views on research results with the lay public, digital affordances also allow researchers to invite the public to get involved in research processes, either by engaging the public in searching, collecting and classifying data (citizen science projects) or by supporting researchers' work through individual donations (crowdfunding proposals). Both citizen science projects and crowdfunding proposals illustrate how emerging genres respond to social exigences (in this case, the democratization of science agenda) and how new scholarly practices are socially situated. In the following sections, we discuss the democratization of science through digital genres that involve both academic knowledge communication and collaboration in scientific research processes.

8.2 Digital Genres and the Participation of the Public

Both citizen science projects and crowdfunding proposals involve the collaboration of the public in research processes along with scientists, either by searching, analyzing and/or classifying data or by supporting researchers through donations, that is, individual citizens' micropatronage. From a (rhetorical) genre perspective, one social exigence that accounts for the emergence of these two genres is the democratization of science agenda, the current movement toward making science more accessible and participatory to broader publics (Follett & Strezov, 2015;

Kullenberg & Kasperowski, 2016). A further and related social exigence is to be found in the so-called accountability of scientific research. Researchers are increasingly engaging in communication practices that involve the participation of the public for reasons of trust and accountability of scientific research and for social impact (Pérez-Llantada, 2021b; Pham, 2016; Prem *et al.*, 2016).

Other arguments support the significance of the democratization of science and the value of getting citizens involved in research processes in various ways. It has been argued that through close collaboration between researchers and the broad public, a more comprehensive view of scientific advances can be achieved (Reid, 2017b, 2019; Reid & Anson, 2019). The growth of published peer-reviewed articles on citizen science projects (and on methods to successfully conduct citizen science research) in the Web of Science and Scopus databases in a decade span (Follett & Strezov, 2015) attests to the increasing popularity of these projects. Increasing popularity also applies to crowdfunding proposals (Mehlenbacher, 2017, 2019b).

These two types of projects result from sociohistorical, situated factors, above all open science policies that have given a prominent role to these new communication demands and promoted this type of communication because of the value of its social and educational reach. In other words, new social exigences have prompted scholars' interest in openly sharing their work to these publics and certain new technological affordances have been developed to support these emerging social exigences. It should be noted, though, that the impact of the open science agenda is by no means dramatic. At least to date, researchers' growing interest in these participatory practices proves to be somewhat limited compared, for example, with their interest in disseminating new knowledge through promotion-related genres, above all, journal articles. Drawing on a survey study of 4000 international researchers, Pérez-Llantada (2021c) reports that there seems to be very little involvement on the part of the researchers both in projects involving the participation of the public and in other emerging digital formats and generic innovations such as AudioSlides, webinars, explainers and author videos accompanying journal articles. Academic discipline is also an important variable regarding researchers' engagement in these genres, as it has been shown that these genres are more important in biological and health sciences compared to other disciplinary fields such as those of the social sciences (Paulus & Roberts, 2018; Pérez-Llantada, 2021c).

8.2.1 Citizen science projects

Citizen science projects are a response to one of the streams defined in the public school of open science, an approach to disciplinary knowledge production and communication that allows citizens to play an active role in the research process through the collection and classification of data, review of the literature and the interpretation of results (Fecher & Friesike, 2014; Reid & Anson, 2019). As defined in the *Encyclopedia of National Geographic*,[1] citizen science is 'the practice of public participation and collaboration in scientific research to increase scientific knowledge'. It has several advantages. One is that community-based groups can participate in generating ideas, seek the advice of the researchers in the processes of data collection and interpret the data along with the research group. Another advantage is seen in the diverse profiles of the citizen volunteers. Scholars can target diverse audiences, ranging from interested people and students and educators, to amateur scientists. The UK Concordat for Engaging the Public with Research defines this type of research along similar lines, namely, a type of communication that 'takes the form of collaborative and consultative roles'.[2] To encourage collaboration, researchers draw on the technological affordances and interactivity possibilities of the citizen science web portal, specifically, the discussion forum available in these platforms for active participation of the project volunteers (Kelly & Miller, 2016). Language is also a key resource for researchers seeking to engage citizens. Reid and Anson (2019: 220) stress that citizen science should be written effectively as its goal is not just to communicate research processes but also to get citizens involved in these processes in different ways, either discussing the findings with the project volunteers once the data collection and classification processes are completed (see also Reid, 2017b), or supporting the researchers in garnering funds for conducting the research. Language is also key to appeal to citizens' interests and curiosity and engage them in research processes. Effective writing ensures that the communicative goals of these projects, informing and educating the public on issues of scholarly research, by this means democratizing science, are fulfilled.

8.2.2 Crowdfunding proposals

Crowdfunding proposals have also emerged in response to social exigences, namely, the need to comply with institutional and funding bodies' demands for making science accessible to society in general (i.e. social impact of research) and the need to increase society's trust in scientific research. At the same time, researchers who launch crowdfunding projects are interested in obtaining funding for their research at a time of increasing competition for funding worldwide and limited resources available. These exigences explain why these projects are becoming increasingly popular across the disciplines, ranging from biology, medicine, mathematics, engineering and data science to psychology, social science, economics, art and design and political science, to name a few. They may also explain why these projects involve researchers with diverse academic profiles, above all, junior researchers and postdoctoral research fellows but also experienced researchers and lab heads. The topics of these projects range from local science (e.g. water quality flowing from Little River and Biscayne Canals into Biscayne Bay; hyena diet in the Ngorongoro Conservation Area, Tanzania; or fentanyl prevalence in New York City) to science addressing global challenges, for example, cancer treatment, infant mortality, or discoveries for the coronavirus disease.

In this genre, the public does not take an active role and has no responsibility in the research process as in citizen science projects. They are receivers of the knowledge communicated and, using the technological affordances of the platforms, they can send messages of support to the researchers, discuss outcomes with them and express their views and opinions about the development of the project and its expected results. However, apart from donations, their participation is somewhat limited. They generally send short comments and replies to the researchers' posts, and give feedback in the form of likes and tweets/retweets and shares in social media (Cody *et al.*, 2015; Mehlenbacher, 2019b; Paulus & Roberts, 2018).

8.3 Platforms that Facilitate Public Engagement in Research

Projects involving the participation of the public (both citizen science projects and crowdfunding proposals) are managed through platforms whose features enable such participation.

The platforms for citizen science projects contain shared infrastructures, repositories and interactive resources which researchers, along with citizens-volunteers, use for collecting, classifying, analyzing, interacting and interpreting data. These platforms arrange the information of the project in modular texts using a set navigating menu. For example, in the platform of Experiment.com, the purpose of the text of the first tab of the menu is to give an overview of the project and its intended goals. In the second tab, the researchers provide volunteers with instructions on how to proceed with the data collection and/or how to classify the data. In the third tab, the interactive tools of the platform allow the researchers to share updates about the project and exchange views with their volunteers. Citizen science platforms also promote the use of social media such as Facebook and Twitter to coordinate the volunteers who join the project and provide them with instructions for collecting data as well as to exchange views on the research in progress. These interactions may take place among local volunteers and/or the global community involved in the project, in the case of projects dealing with global issues or encompassing a global (supranational) audience (e.g. in EU-Citizen. Science projects). Other technological resources, such as smartphones, tablets and tools for survey research also enable the volunteering citizens to collaborate with researchers in fieldwork (Chivite, 2017).

In the spirit of democratizing science, the Citizen Science Association,[3] based in the United States, uses a platform that supports this type of projects, offering citizens the opportunity to volunteer for projects that span the disciplines from arts, biology, history, language and literature to physics and social sciences. Similarly, to support the citizen science movement, the EU-Citizen.Science platform offers shareable resources and tools and provides guidelines and recommendations for successfully launching projects and engaging interested citizens in research activities such as collecting, analyzing and discussing data online with researchers. The projects of this platform deal with a wide range of topics, from ecology and the environment, nature and the outdoors, long-term species monitoring, health and medicine, to indigenous culture, social sciences, information and computing science. While some projects may be doable at home, which means that volunteers can collect and classify data from home, others require fieldwork for data collection and classification. Other popular platforms are SciStarter, Zooniverse, BioBlitz and Adventurers and Scientists for Conservation, designed along similar lines. Table 8.1 shows examples of citizen science projects of Zooniverse, a platform that supports researchers from a range of fields. It also gives an overview of the projects' level of engagement with the public (in terms of number of volunteers) and the contributions made by the volunteers (measured in number of classifications of the data collected for the project).

How researchers engage citizens very much relates to the technological affordances of the platforms. As stated, the web interface of Zooniverse projects is modular, which means that the web layout enables the researchers to post the information of the project in the different tabs of the web page menu. Navigating the menu from left to right, the reader first finds the tab 'About', which works as an introduction and includes an overview of the project and general information on its scope and goals, as also stated previously. In the second tab ('Classify'), researchers provide instructions and explain to their volunteers the steps to follow to search, collect and/or classify the data correctly. The third tab ('Talk') is an interactive tool that works like a blog with which researchers interact with their volunteers, exchange views and discuss the data that have been collected. The last tab ('Collect') acts as a repository of the collected data. The 'About' tab also splits the information of the project into modules, including sub-tabs that, from left to right, provide information about the research goals, the researchers' bionotes and photographs, evidence of the researchers' credentials (e.g. selected publications) and the educational goals of the project.

In crowdfunding proposals, we also find that technological affordances are harnessed by researchers to set up new research with the help of the project backers' micropatronage (Fecher & Friesike, 2014; Paulus & Roberts, 2018). While citizen projects involve 'volunteer

Table 8.1 Sample of citizen science projects in Zooniverse

Project	Accompanying description	Field	No. of volunteers	No. of classifications	Location
Mapping historic skies	Help us identify constellations in celestial maps from the Adler Planetarium's collection.	Arts	2399	35,568	Chicago, IL
Nest Quest Go!: Wood Duck	Explore the curious life cycle of ducks that use nest boxes.	Biology	254	98,689	New York
Notes from Nature - NYBG	Discover historic specimens and drive global plant research at one of the world's largest herbaria!	Climate	384	105,228	New York
Transcribe Cooper	Help Uncover the Hidden Aspects of Anna Julia Cooper's Life and Writing	History	533	6,627	Washington, DC
Maturity of Baby Sounds	We need your help to classify some very short recordings of babies' speech sounds. This will help us to better understand the very first stages of language learning!	Language	586	26,994	Rosario, Argentina
Worlds of Wonder	Explore the worlds of wonder seen through the microscope. Help us identify and classify nineteenth-century microscopy illustrations.	Literature	1964	82,905	Maastricht, The Netherlands
Etch A Cell – Powerhouse Hunt	Etch A Cell is back and this time we're searching for mitochondria - the powerhouses of your cells.	Medicine	1477	44,428	London, UK
Expedition Arctic Botany	Transcribing the National Herbarium of Canada's Arctic Collection.	Nature	837	19,861	Ottawa, Canada
Radio Galaxy Zoo: LOFAR	Help astronomers locate and identify supermassive black holes and star forming galaxies!	Physics	2592	230,202	Leiden, The Netherlands
Power to the People	Help us find rural homes in Africa to expand electrical access.	Social science	529	18,988	Oxford, UK
Star Notes	Help us transcribe the groundbreaking work of early women astronomers!	Space	1678	76,242	Cambridge, MA

Source: Retrieved from https://www.zooniverse.org/projects (last accessed March 17, 2020).

engagement as a way to extend methods for acquiring, integrating and modelling massive quantities of diverse data' (Scheliga *et al.*, 2016: 518), crowdfunding project backers simply support the project team and the project development through donations, which make it necessary to include elements that prompt prospective backers to donate (e.g. the budget overview, including a rationale for the expenses, or endorsement by researchers). Examples of platforms enabling public participation in research processes through micropatronage are Consano and MedStartr (for medical research projects and medical and healthcare innovation projects, respectively), and Kickstarter and Experiment.com (for projects across different disciplinary fields). These platforms exhibit similar ways of displaying the Overview text of the project, namely, by arranging the information under different headings. As seen in Table 8.2, these headings reflect the genre's response to the social exigences of transparency, accountability of science and social impact (e.g. main benefits).

Like citizen science projects, crowdfunding proposals also break down the information of each project using a menu with different tabs. For example, in Experiment.com, the homepage includes tabs for (from left to right) Overview, Methods, Lab Notes and Discussion (see Figure 8.1). In this platform, in addition to the project overview, the researchers include information on their research team, a photograph of the team, details of the project timeline and an explanation of the funding goals. They also provide a summary of the methods and procedures they have designed to realize the project goals, post updates for their backers using the 'Lab notes' tab and interact with their followers using the 'Discussion' tab. Splitting the project information into these different sections may make it easier for the researchers to write the texts and allocate contents to each section/tab. In addition, it may also facilitate the cognitive processing of the information on the part of the public. This approach seems suitable for this genre, especially because one of its goals is to increase citizens' scientific literacy (Bonney *et al.*, 2016; Mehlenbacher & Mehlenbacher, 2019).

Table 8.2 Headings of the projects' Overview in crowdfunding science platforms

Consano	Experiment.com	MedStartr	Precipita
Summary Why is this important? Who will benefit? Budget	About this project What is the context of this research? What is the significance of this project? What are the goals of the project? Budget Endorsed by Project timeline Meet the team	About our project About our solution About our team About our company How we help patients How we help physicians How we help partners Innovation details	About the project Why is this project original? Where are we? Who will benefit? Funding targets

Figure 8.1 Homepage of an Experiment.com project launched by Rodolfo Garcia-Contreras and Jesús Guillermo Jiménez-Cortés (Project doi: 10.18258/10486; retrieved from https://experiment.com/projects/are-current-antibacterial-treatments-resulting-in-the-evolution-of-more-multidrug-resistant-bacteria – last accessed 27 February 2021)

8.3.1 Language(s) for engaging the public

The technological platforms available for communicating disciplinary knowledge beyond expert audiences and engaging broader publics in research processes are widely used by multilingual researchers. When the platform is supported institutionally, for example by a national funding agency or by a research institution or organization, the language used is the national language. In some platforms the information can be accessed in more than one language. As an example, Precipita, the platform of the Spanish Foundation for Science and Technology for crowdfunding Spanish science projects, is multilingual, that is, it has both a Spanish and an English version for each project. The most popular international platforms are English medium, given their international reach. As seen on their web portals, the majority of the project launchers are multilingual researchers from diverse linguacultural backgrounds and geographic locations, from Africa and South America to Asian countries. English is therefore used as a shared lingua franca. Yet, researchers may also occasionally use their native language to interact with their backers if they share the same language (usually, the language of the local setting). Hyperlinking affordances also make these platforms multilingual sites for research communication. For example, in English-medium crowdfunding science platforms, the researchers sometimes provide hyperlinks to their publications, professional websites and patents and while some are in English, others are in languages other than English (Pérez-Llantada, 2021b). Additionally, researchers can also draw on the multimodal affordances of the platforms and some include explanatory videos in two or more languages. In these videos, the verbal message of the video is conveyed in English and the subtitles are displayed in the local/national language(s).

As to how language(s) is/are deployed in these emerging genres, researchers mainly use an informal style that creates proximity with their volunteers/backers (Pérez-Llantada, 2021a). They also keep the use of expert jargon to a minimum to facilitate comprehension of specialized contents. Disciplinary knowledge is conveyed clearly and effectively through paraphrasing strategies, as is also the case of science popularizations (Gotti, 2014) and digital genres such as TED Talks (Caliendo, 2014; Scotto di Carlo, 2014). In addition, both citizen science projects and crowdfunding proposals use language persuasively to encourage collaboration and to prompt donation, respectively (Paulus & Roberts, 2018; Pérez-Llantada, 2021b).

8.3.2 Multimodality

Multimodality is an important feature of genres for engaging the public in research processes, both citizen science projects and crowdfunding proposals. In the former projects, visuals perform several functions. For example, the visuals used in the Overview section/tab, where researchers introduce the project and explain the relevance of their research goals, illustrate and contextualize the project. Photographs show the object/phenomenon investigated and/or the researchers conducting fieldwork. In the Classify section/tab, photographs, videos and audiovisual tutorials perform a primarily didactic function, that is, they illustrate the objects that the volunteers have to search, collect and/or classify and what procedures they should follow to do so. In crowdfunding proposals, the Overview section/tab includes either an image or an embedded video pitch. The image conveys the main topic of the project or the object of study, as shown in Figure 8.1. At other times, photographs and videos relate the problem that the researchers need to solve, or show the researchers doing fieldwork once the project has started, as we will see in the case study analyzed later.

To understand different systems for meaning making in genres that combine multiple semiotic modes, Jones and Hafner (2012: 60-63) propose a framework for multimodal analysis that considers how image and text are combined to make meaning. If we apply Jones and Hafner's (2012) framework, we see that in both types of projects, although both text and image tend to be equally important, textual information is more prominent for achieving informational goals. Because the images are more immediate than text, they create greater persuasive appeal. It can therefore be said that although in general there is no dominant mode, the position of the images in some of the sections/tabs makes images more salient, for example, in the About section/tab of citizen science projects and in the Overview section/tab of crowdfunding proposals. In the Classify section/tab of citizen science projects, visuals are also more salient than text for pedagogical purposes, while in the Lab notes tab of crowdfunding proposals, the verbal mode prevails over the visual mode and visuals are mainly illustrative. Regarding image-text interaction, the messages conveyed by both modes are at times concurrent, that is, the messages are the same and reinforce each other and, at other times, complementary, when the visual mode fills in the details of the verbal mode.

To show how researchers nowadays use the technological affordances of the platform to inform about their project, raise public interest in issues of science and raise the necessary funding to set up a research project, and that language and multimodal resources are key resources to achieve those goals, in the following section we explore a crowdfunding project.

8.4 Research Case Study: Prompting Micropatronage in a Crowdfunding Proposal

In this section, we analyze a successfully funded crowdfunding science project[4] launched by Mario Williams and Jacquelyn Gill, a graduate student and researcher, and an assistant professor (at the time the project was launched) at the University of Maine Climate Change Institute and School of Biology and Ecology. The aim of the study is to examine how the information about their project, '10,000 Years of Climate and Environmental Changes in Jamaica, a Biodiverse Tropical Island', is organized informationally, and how language and multimodal resources are used, to achieve the purpose of the genre. The first researcher, Mario Williams, originally from Jamaica, launched the project for both personal and professional reasons. On the personal side, he notes, 'I have always wanted to contribute meaningfully to the development of this beautiful island that I call home. [...] a special opportunity to fulfil this lifelong dream'. On the professional side, he writes, 'It is my hope that scientific advances in biodiversity conservation will enable us to protect the natural environments that we rely on, so that future generations will be able to appreciate and enjoy nature as we do today' (available in the 'Meet the Team' section of the project

Overview page https://experiment.com/projects/10-000-years-of-climate-and-environmental-changes-in-jamaica-a-biodiverse-tropical-island). The second researcher, Jacquelyn Gill, leads the Biodiversity and Environment Across Space and Time Lab at the University of Maine and, as her bio statement notes, is 'passionate about science communication and diversity in STEM'. She mentions her interest in sharing her research beyond the expert community: 'I started blogging and tweeting about science as ways to make science more accessible'. They launched the project in 2017, garnered 173 financial backers and successfully raised the funding needed to carry it out (an amount of $10,043).

Specifically, we address these questions: (i) How do the researchers use language resources in the different sections/tabs of the platform web page?; and (ii) What multimodal resources do they use along with the verbal mode? For the analysis of language we use genre and discourse analysis (Hyland, 2010; Swales, 1990, 2004) and for the analysis of multisemiotic resources we use the multimodality analysis framework proposed by Jones and Hafner (2012).

As we explained earlier, in the Experiment platform all project web pages share the same (fixed) menu tabs (Overview, Methods, Lab Notes and Discussion). In the Overview tab, the information of the project is arranged into sections by means of headings ('About this project', 'What is the context of the research?', 'What is the significance of the project?', 'What are the goals of the project?', 'Budget', 'Endorsed by', 'Timeline' and 'Meet the team'). Table 8.3 lists the functions of these parts in the crowdfunding project analyzed (see also Mehlenbacher, 2017, 2019b for further discussion).

Table 8.3 Overview section of the crowdfunding proposal

Parts of the overview section	Functions
About this project	Sets the research scene and provides an overview of the project
What is the context of this research?	Establishes the 'research territory' and creates 'a research niche' (Swales, 1990)
What is the significance of this project?	States gap in knowledge in real world; leads to motivation for crowdfunding
What are the goals of the project?	States the purpose(s) for requesting the funding
Budget	Justifies the expenses and explains how the funds raised will be spent
Endorsed by	Shows other researchers' support for the project and helps the project launchers to claim competence
Project timeline	Provides a timeline for the completion of the different project tasks
Meet the team	Provides a biography of the members of the team and their motivations to launch the project

This rhetorical organization of the project's text aims to inform, raise interest and persuade the interested public. For example, in the 'About this project' tab, the project launchers first situate their research geographically and then raise the ecological problem that they have identified and aim to address by conducting the project ('Jamaica is a highly biodiverse island in the Caribbean, but its ecological resources are threatened by climate and land-use changes'). Afterwards, they formulate the rationale for their project ('We need a better understanding of how the island's species responded to past climate and human impacts') and do so using the first-person plural pronoun 'we', a language resource that may work as self-representation (Swales, 1990), as well as interpersonally, i.e. to achieve persuasive goals, if considered to be an inclusive pronoun that textually constructs 'both the writer and the reader as people with similar understandings and goals' (Hyland, 2010: 116). Finally, the researchers explain what the funding obtained will be allotted to ('Funds raised through this campaign will help us collect sediment cores to develop a 10,000-year long environmental record of vegetation, fire, climate, and human impacts in Jamaica'). Under the heading 'What is the context of this research?', language is mainly used to inform the project backers further. The researchers highlight the value of the physical environment and explicitly note aspects not identified in the previous literature to justify their project ('The island has several sites that are designated as internationally important conservation areas under global treaties. Few studies have investigated past climate and environmental changes in the Caribbean islands, and of these, Jamaica is one of the least understood'). Rhetorically, these statements pave the way for informing backers about the project impact ('This research will fill an important knowledge gap in Jamaica's environmental history, and the findings will create new possibilities for both ecological and archaeological research on the island') under the heading 'What is the significance of this project?'. Finally, under the heading 'What are the goals of the project?', the researchers explain in non-technical language what they will do with the funding ('Funds raised through this campaign will allow us to travel to Jamaica to collect lake sediment cores. [...] We will use these records to analyze the fire, vegetation, climate ...'). As Mehlenbacher (2019b) states, this sequential information organization is an effective way to fulfil two communicative goals in this genre: informing clearly about the project and persuading the public to donate.

Language is used in specific ways to achieve these two goals. For example, to claim significance, the researchers use evaluative nouns and adjectives such as 'important', 'significant benefits', 'new possibilities', as well as the modal verb of certainty, 'will', to explain how their expected scientific contribution will address the stated concern. To convey credibility and give relevance to the object(s) investigated, the researchers choose an impersonal tone expressed by inanimate subjects such as 'this research', 'the findings', 'project'. In doing so, the researchers' account of the benefits of the project outcomes for the ecological maintenance and preservation of natural resources reveals a careful 'narrative crafting' (Paulus & Roberts, 2018) that creates a persuasive appeal to their backers.

In the 'Methods' section/tab, language is mainly used for informational goals. In this section, the researchers include a summary of methodological procedures, a brief statement of the challenges they expect to encounter in the research process and a pre-analysis plan or schedule for implementing the protocol. As seen in Example 8.1, although the researchers employ specialized jargon to refer to methods and materials, they report in the first person ('we'), which makes the account personal and informative enough for a non-expert audience. The modal 'will' conveys immediacy of action and intentionality, or willingness to move the project forward.

> **Example 8.1** We will collect and analyze a sediment core from Wallywash Great Pond which is located in southwestern Jamaica, and is the island's largest lake. We will then examine the sedimentary record of pollen, charcoal, ostracod δ18O, and Sporormiella at Wallywash Great Pond to develop a high resolution paleoenvironmental history of vegetation, fire, climate, and human impacts in Jamaica spanning > 10 ka.

What is interesting from this section is that while the researchers use verbs such as 'collect', 'analyze', 'examine', to describe research actions, and the modal 'will' that conveys intentionality and do so in a conversational style (e.g. as in 'is the island's largest lake'), at times they provide technical descriptions with specialist jargon that they nonetheless exemplify and clarify to their backers in non-technical language. Example 8.2 shows the merge of specialized discourse and pedagogical discourse (the latter highlighted in italics).

> **Example 8.2** $\Delta^{18}O$ values measured from ostracod shells reflect the precipitation-evaporation regime of lakes, *because ostracods incorporate oxygen from their host lake waters during growth.* The lighter oxygen isotope (^{16}O) preferentially evaporates during periods of high evaporation; *therefore, changes in the isotopic composition of ostracod shells in lake sediments can tell us about relative moisture availability in the surrounding lake area.*

A formal register is also used when the researchers speculate about the validity of the criteria for data collection and do so in a different style, not accessible to lay publics ('The concomitant fulfillment of multiple criteria before the established date for first human settlement of the island would strongly suggest that this date should be revised'), which resembles the expression of limitations in journal article writing.

'Lab notes' and 'Discussion' allow the researchers to establish a dialogue with their backers. In this case study, Mario uses these affordances to write two posts in the 'Lab notes' section, one giving an update on the start of the project (the researchers' trip to Jamaica) and the other one commenting on what he is learning during this trip. In both posts, the researcher uses language to inform backers, educate them and increase their enthusiasm in science. By way of illustration, in the Lab notes section/tab the researcher uses a colloquial (conversational) style to report to backers, but also exclamatives in the titles of their posts ('Major Fieldwork Update! Video included!' and 'Paleoecological training in the BEAST Lab!') to persuade backers to read the posts. In both posts, Mario uses informal opening and closing salutations ('Hi everyone' and 'Cheers') and his first-person pronoun narrative on the team's trip and arrival to Jamaica is both informative ('My trip to Jamaica lasted from February 2–10, 2018. My travel partner and trusted field assistant was...') and anecdotal ('Andrea and I were relieved to have arrived safely in Jamaica – after our flight had been cancelled coming out of Bangor Airport in Maine – and also to hear that all of our checked luggage and field equipment had arrived on the island'). By this means, he conveys credibility and constructs proximity with their backers, who respond to the post in the form of views, likes and comments. This conversational style adheres to the guidelines of the Experiment platform for writing Lab notes and Discussion – 'the same style you would use when talking to another human in real life, conversational'. Moreover, Mario uses language for interpersonal goals, specifically, to formally acknowledge other researchers' support for their assistance and the support of funding institutions. He also thanks their backers profusely for their support using evaluative, colloquial adjectives ('195 amazing and awesome donors via Experiment.com!'). In the second lab note, the post starts with an emotive note and Mario again uses exclamatives and second-person pronoun references to thank their backers in a very friendly way ('Thank you very much for your contributions to our project. Your love for science and generous support will ultimately make this work possible!'). In stark contrast, this is followed by a detailed reporting of research processes in which the researcher combines specialized terminology and research-process verbs with the use of a language accessible to non-expert readers, thus reflecting the accountability and educational social exigences of this genre.

Example 8.3 We visited Pushaw Lake, a large and shallow lake located approximately 15 minutes away from the UMaine campus in Orono, and successfully collected several sediment cores using a Livingstone piston corer. We observed different sediment layers within the collected cores, ranging from dark-brown gyttja with high levels of organic matter and plant macrofossils to bluish-gray, fine-grained glaciomarine clays. These

> sediment layers reflect the varying climatic conditions that existed in Maine over the past >12,000 years, and provide a long-term record that can be used to understand how Maine's landscapes changed during the most recent glacial-interglacial transition. We will assess a similar lacustrine paleorecord for Jamaica to better understand long-term linkages between climate, humans and the environment on the island.

Interestingly, the researcher also shares with their backers his life experience and the skills gained during the trip ('Over the past few weeks, I have been working with Jacquelyn to develop both my lab and fieldwork skills in preparation for our investigation into Jamaica's ecological past. I am receiving training in a variety of research techniques...', 'I have also been developing my pollen identification skills in the Paleoecology Lab at the UMaine Climate Change Institute'). His informal, first-person narrative contains easy-to-grasp definitions of the objects of research and use of similes, for didactic goals ('For example, pine (genus Pinus) pollen grains are bisaccate and have two bladders attached to a central body, which makes them resemble Mickey Mouse ears!') and paraphrases clarifying specialized terms by giving examples ('Grasses [family Poaceae], including major cereal crops such as rice, corn and wheat, generally produce large, spherical, monoporate pollen, *which means that* there is a single pore on the surface of each pollen grain', emphasis added). Moreover, the researcher even draws on translanguaging to prompt donation while revealing his linguacultural background ('Increasing your donation, as all contributions you make no matter the size will have an immense impact on the eventual outcomes of this project! As Jamaicans would say in our traditional dialect, "Every mickle mek ah muckle", which means that "Every bit counts!"').

Finally, the Discussion section/tab acts as an online forum where the backers can support the researchers and express their views. Example 8.4 presents a thread of comments in which the project backers support the research team and congratulate them for having raised the funds needed to conduct the project. The language used is informal, as deduced from the use of expletives, first-/second-person pronouns, colloquial expressions, subject ellipses and contracted forms. Further, the comments, numbered in the order in which they were made, convey the followers' excitement about science, sharing relevant science concerns, funding them and the impact of scientific achievements.

Example 8.4 Comment thread

(1) So excited to see it's fully funded! Congrats!
(2) This is great, good luck!!
(3) Congratulations!!
(4) I don't have much to spare, but I hope this helps, and I want to see this thing go through.

(5) This looks like such a great project!
(6) This has all the markings of ground-breaking research!
(7) Good luck. Hope you make your goal and get to do this important work.
(8) Awesome work, Mario! Thank you for doing such important research - I'm sure the results of this work will be very impactful for research efforts related to Jamaica for many years to come!

This analysis also illustrates how while the technological platforms supporting public engagement in science impose constraints (fixed templates, text modularity and fixed interactivity tools, limited number of words, detailed writing guidelines), the researchers make creative use of language and multimodal resources. For example, in the web page of the project, the researchers include a visually appealing video pitch in which Mario introduces himself, shows some beautiful shots of a Jamaican beach, the researchers' fieldwork and uses an informal style to invite the public to fund the project (see Figure 8.2). The video pitch also includes

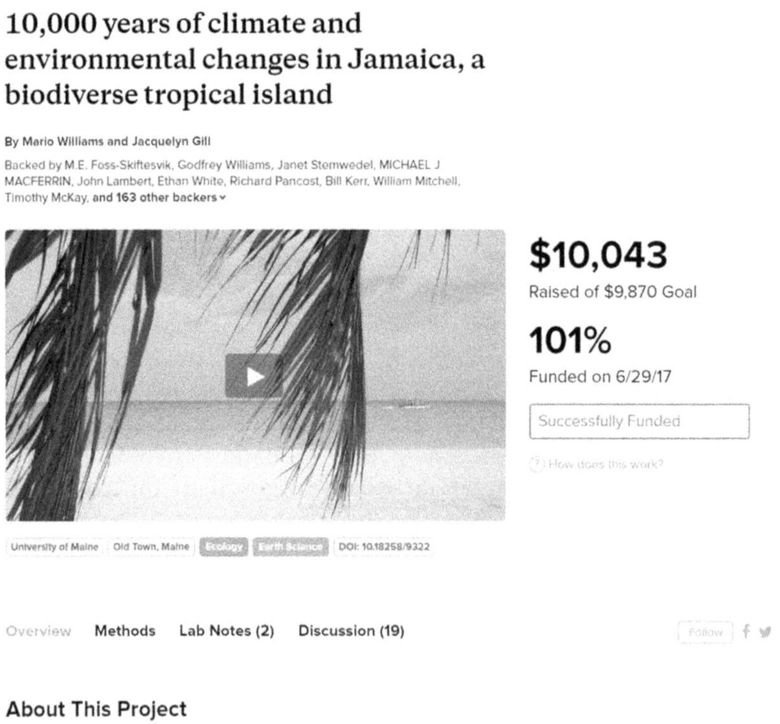

Figure 8.2 Homepage of the project '10,000 Years of Climate and Environmental Changes in Jamaica, A Biodiverse Tropical Island', Mario Williams and Jacquelyn Gill (doi: 10.18258/9322; https://experiment.com/projects/10-000-years-of-climate-and-environmental-changes-in-jamaica-a-biodiverse-tropical-island – accessed 16 May 2021)

some shots of the researcher working in his office with his computer, which help to construct a professional persona. This is followed by shots of several animal and plant species that highlight the importance of preserving biodiversity, one of the main issues highlighted in the Overview text. The video pitch closes with some shots of the researcher in his lab explaining his plans to visit the island to do fieldwork if the necessary funding is raised. This multimodal narrative fulfils several purposes: to introduce the researcher as a person; to summarize and highlight the main ideas of the project overview text; to give credibility to the researcher's work; to explain the goals for the project; and, more broadly, to inform and sensitize the public about the importance of science addressing environmental issues.

In the Lab notes tab, the predominant mode is the verbal mode. The visuals are at times concurrent and at other times complementary. In Lab note 1, the researcher includes a hyperlinked YouTube video pitch with clips and images that summarize their trip to Jamaica, starting with the researchers' arrival on the island, continuing with instances of fieldwork showing how they collected data, and closing with an image of their plane leaving the island. Mario encourages their backers to watch it using exclamatives ('Please take a few minutes to watch the video .... Enjoy!!'). Image-text interaction in this case is concurrent, as both text and video pitch convey the same message though using different modes. In Lab note 2, some of the photographs inserted in the main text are complementary to the message conveyed in the text. But because images are polysemous, the selected also serve to convey (i) the researchers' love and enthusiasm for science and the importance they give to scientific research; (ii) the topological details of the physical location where fieldwork takes place; and (iii) Mario's personal experience as a junior researcher (the way he is learning the methodologies and developing research skills) (Figure 8.3a and 8.3b). In short, the visual and verbal messages are the same and reinforce each other.

The other photographs inserted in the main text of Lab note 2 (Figure 8.4a and 8.4b) further illustrate what the researchers report in the text and thus have a pedagogical goal, illustrating the researchers' instruments and methodological procedures 'to assess when humans first settled Jamaica and the response of the island's vegetation to past natural and human-caused disturbances'. Again, the verbal-visual combination plays a crucial role in constructing the researchers' authority and credibility (Hyland, 2012), in conveying professionalism while making visible (and shareable with their backers) a personal identity and in increasing the public's scientific literacy.

In sum, this exploratory analysis has illustrated the affordances that crowdfunding platforms offer researchers worldwide to engage the public in research processes, share scientific advances, build trust in science and increase the public's interest in science. From a genre

(a)

Behold the wonder of the lake sediment core!

(b)

Two thumbs up for science!

Figures 8.3 Photographs of researchers in Lab note 2

analytical perspective, the case study has foregrounded how crowdfunding proposals fall within the category of parascientific genres insofar as they adopt organizational and linguistic features of existing genres while exhibiting distinct features as well, and should thus be 'taken as a typified

(a)

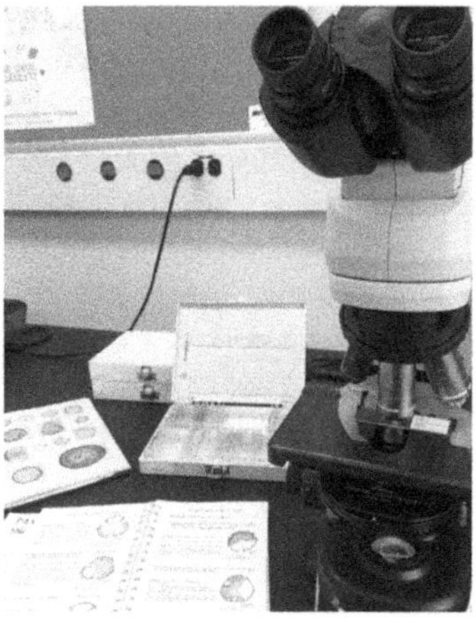

Here, a palynologist will sit quietly and ponder the mysteries of past worlds

(b)

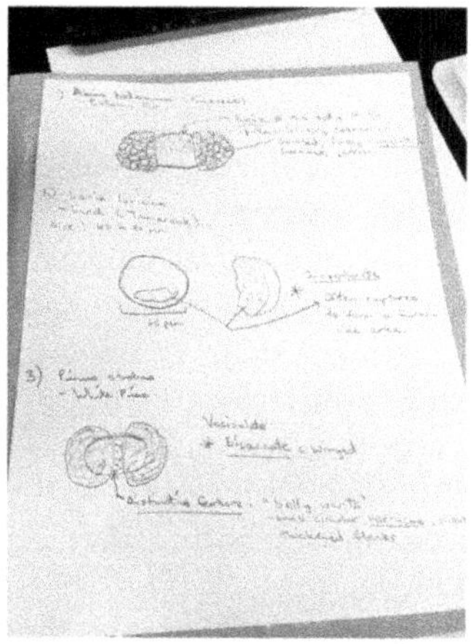

Building a personal pollen guide with sketches and descriptive notes for each observed grain

Figures 8.4 Photographs of research settings in Lab note 2

meaningful action that is recurrent and predictable *to a certain extent*' (Kelly & Miller, 2016: 223) (emphasis added).

8.5 Concluding Remarks

In this chapter, we have illustrated how new technologies afford researchers the opportunity to communicate their work with diversified, non-specialist audiences and engage with them in research processes. We have shown how rapidly changing technologies represent a turning point for disciplinary knowledge dissemination insofar as they enable researchers to communicate science beyond the expert community and engage the public in research in different ways. Both an informal discourse style and multimodal affordances (photographs, video pitches and animations) are combined with the verbal mode to create different rhetorical effects. We have also seen how researchers tell and sell their science through the construction of both a personal and a professional identity, which can be very different from the one they construct in peer-to-peer scholarly communication.

Notwithstanding these technological affordances, broader issues emerge from this exploratory study that would need addressing in future research. As we have argued elsewhere (Pérez-Llantada, 2021b), the first is whether these new practices have value in existing research assessment systems (Lillis & Curry, 2010). The second, whether researchers will devote time and effort in engaging in these emerging practices if they do not obtain personal benefits (e.g. academic recognition and promotion) as compared with spending their time in writing 'countable' genres (Tusting, 2018). We might expect that they would rather opt for research outlets that are valued for promotion and recognition purposes in a 'publish or perish' research world, for example, in writing traditional forms of research communication such as journal articles. Even if the genres that we have discussed in this chapter embody the principles of open science, the existing systems of rewards and regulations do not yet value them as much as other research outlets such as journal articles, for example, to measure research impact.

Another issue, once again, is access to these technological affordances. It is possible that researchers in the periphery of academic knowledge production may lack the internet facilities to engage in these emerging practices if in the short or long run they become mainstream. A third issue concerns digital multimodal composing skills. Some researchers may find it difficult to engage in effective multimodal genre composing (see Hafner & Ho, 2020, for a detailed description) and may not have access or opportunities for formal instruction to learn these skills. These are some socioeconomic and contextual aspects underpinning the genres that we have explored in this chapter and are areas that need further exploration and consideration.

Notes

(1) https://www.nationalgeographic.org/encyclopedia/citizen-science.
(2) See https://re.ukri.org/documents/hefce-documents/concordat-for-engaging-the-public-with-research, p. 4.
(3) https://citizenscience.org/.
(4) In this chapter we critically comment on a single exemplar of the genre so that the reader best captures how language and semiotic resources intersect in this genre type.

9 'Showing' Research through Audiovisual Genres

9.1 Introduction

Previous chapters have discussed how digital genres make it possible for researchers to engage in new written communication practices, reach wider audiences and develop new relationships with various stakeholders. The incorporation of digital genres to the repertoire of genres used by scientists and the shift from page to screen (Snyder, 1998) have also increased the range of semiotic resources for research practices and academic communication. An important affordance of the online environment is the possibility of combining a variety of modes (verbal, aural, visual) to convey meaning. New technologies have made the use of a multiplicity of modes easy and natural, and users can easily combine the modes that are better aligned with their communicative needs (Kress, 2003). The digital environment facilitates the complex interweaving of various modes to represent scientific knowledge, and also the connection between different media on a single platform and their linking to elements outside the platform. Various media types can be linked to another, embedded, repurposed and recontextualized in a different platform (Adami, 2014).

As explained in Chapter 2, many digital genres for knowledge communication involve the remediation, 'the formal logic by which new media refashion prior media forms' (Bolter & Grusin, 1999: 273), of printed academic genres in the digital medium. This remediation usually involves harnessing the multimodality affordances of the new medium, and combining verbal and visual resources that enable authors to engage in 'showing' rather than 'telling'. Online videos in particular are an effective tool for communicating with peers, for addressing the reproducibility crisis by providing detailed descriptions of methods (Hafner, 2018) and for promoting researchers' work (Luzón, 2019b). Researchers also use videos to make their research understandable and engaging for a diversified audience consisting of peers, researchers in other disciplines and the interested public (Luzón, 2019b).

In this chapter, we first review studies on digital (audio)visual genres used to 'show' research. We discuss different types of (audio)visual abstracts, the video methods article (VMA) and stand-alone videos to communicate with a wider audience. In the second part, we illustrate how multimodal resources are orchestrated in online videos by presenting the results of a study of digital videos produced by research groups to disseminate their research.

9.2 (Audio)visual Genres for Knowledge Communication Online

All the genres that we discuss in this section, except for the graphical abstract (GA), are video genres. Videos are highly suitable for showing research processes to peers (see Hafner, 2018) and disseminating knowledge to the semi-specialized or interested public (Erviti & Stengler, 2016; Kousha *et al.*, 2012), because moving images facilitate comprehension of concepts and processes that are difficult to explain with a written text or with a static image (Pasquali, 2007); in addition, and importantly, they 'can transmit emotions that may involve the viewer and promote engagement with scientific issues for a wide group of citizens' (León & Bourk, 2018: 5).

9.2.1 (Audio)visual abstracts

We situate our discussion of (audio)visual abstracts in the context of generic innovation in online scholarly journals. As Mehlenbacher and Mehlenbacher (2019) remark, the research article (RA) has remained highly stabilized when moving online because it responds to the same primary rhetorical exigence as does the printed article: to communicate research findings to experts. However, Web 2.0 technological developments have prompted the emergence of new genres that exist in a constellation around the online article to respond to new rhetorical needs. Various types of visual abstracts have emerged that increase the visibility of the research reported in the article or make it accessible to diverse audiences, for example, GAs (Hendges & Florek, 2019), video abstracts (Plastina, 2017; Spicer, 2014) and AudioSlides (Yang, 2017). These genres borrow features of their traditional generic antecedent, the article abstract, while benefiting from the affordances of the visual mode. According to Brunner (2016), GAs, video abstracts and AudioSlides share three features that help the dissemination of knowledge online: they are easily and freely accessible by everybody; they are short, which helps users to evaluate their relevance for their purposes; and their credibility is easily assessed because they are linked to a reputable journal.

The GA is part of Elsevier's 'The Article of the Future' project, although it did not originate in the digital environment. The first GA was published in 1976, in the chemistry journal *Angewandte Chemie* (Hendges & Florek, 2019; *Nature Chemistry*, 2011). As at that time the journal was multilingual, the origin of GAs seems to be related to the visual nature of chemistry and the role of symbolic chemical language to facilitate comprehension for readers from various linguistic backgrounds (Lane *et al.*, 2015). On the Elsevier website, the GA is defined as a 'visual summary of the main findings of the article'. Studies on GAs concur that this is a promotional genre (Hendges & Florek, 2019; Lane *et al.*, 2015), but tend to present it as still in an experimental phase. Studies do not present conclusive results regarding the visual and linguistic features of this genre, or its relation with the verbal abstract and the other visuals in articles (Hendges & Florek, 2019).

Although the purpose of the GA is to attract the readers' attention and persuade them to read the paper, there is a great variety in content and form (e.g. charts, diagrams) (Hendges & Florek, 2019). This genre has not yet found a common form or mission, and the reason may lie in what one of the participants in Reid's (2017a) study of researchers' multimodal composing felt about this digital innovation. He considered that the GA did not fit with scientific reading practices since it was separated from contextualizing verbal resources and thus did not function as a 'semiotic hybrid' (see Lemke, 1998), served no scientific communicative purpose and conflicted with other already typified parts of the article (e.g. the abstract), which scientists would read instead of the GA. More research is therefore necessary to assess the uptake of the genre and potential users' responses to see whether the genre actually responds to a rhetorical exigence or might be an unnecessary innovation.

Another promotional genre increasingly used to publicize research papers is the video abstract. Some journals (e.g. the *New Journal of Physics*, *BMJ*, journals published by Cell Press) provide short video summaries of their published papers, either embedded on the journal website as a complement to papers or in their YouTube channel. This genre emerged as a result of the push for open-access to academic output and the need to promote research in the current competitive academic context, as well as to take advantage of the visualization possibilities of the internet (Berkowitz, 2013). The video abstract has clear benefits for journals and authors (Plastina, 2017; Spicer, 2014). Publishing video abstracts in open access (which can be linked to, reused in other pages and shared via social media) attracts readers to the published paper, and increases the visibility of the research and its citation rates. Video abstracts also help authors to communicate complex concepts, to address the audience in a personal and engaging style, and thus contribute to reaching a wider audience (Berkowitz, 2013; Spicer, 2014). For some scientists in disciplines such as the health sciences or physics, the main advantage of video abstracts is their potential to present complex information visually and thus overcome some of the barriers of traditional publishing. Given the prominent role of this genre in some journals and the benefits reported by publishers and authors, scholars are exploring its generic features (Liu,

2019). Plastina (2017), for instance, examines how the written abstract is rearticulated multimodally in the video abstract by comparing the move structure of the written RA abstract and that of video abstracts. The results attest to genre innovation and flexibility when pre-digital genres move online. They show that video abstracts include steps not present in traditional written abstracts in order to achieve a stronger promotional effect, that visual and verbal elements are combined to increase conceptual clarity and that the video abstracts blend features of the written abstracts and other genres, such as the conference presentation.

Although video abstracts are typically published on journal websites next to the papers they summarize, authors can use this video format to promote their work on other platforms. One such platform is *Latest Thinking*, a project self-described as 'an Open Access Video Journal' which hosts videos summarizing peer-reviewed academic publications in all academic fields. The journal publishes interview-based videos (about 10 minutes long), divided into five sections (Research Question, Method, Findings, Relevance, Outlook) (see Figure 9.1), where the authors present a summary of the published research, including the significance of the results. Together with the video, the platform offers other types of information promoting the author (e.g. a link to the publication[s] on which the video is based, academic information about the author and their publications, information on their institution) (see Figure 9.1). Although

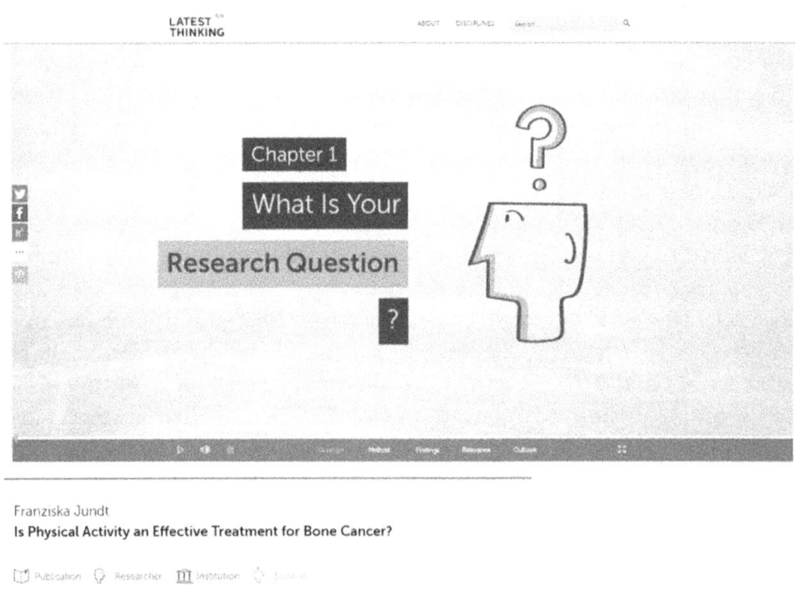

Figure 9.1 Example of a video on the platform *Latest Thinking* (https://lt.org/publication/physical-activity-effective-treatment-bone-cancer) by Jundt (Licensed under Creative Commons Attribution 4.0 International)

Latest Thinking produces the videos, researchers can embed and share them in their institutional or personal website. As the *Latest Thinking* journal is intended to improve the impact of research and to provide a service for researchers who want to increase their visibility, the videos have clear promotional purposes. This project shows how the genre of video abstracts, before becoming stabilized, is evolving to meet specific needs, illustrating the dynamic nature of remediated genres (see Pérez-Llantada, 2021b), or genres that have migrated online. It also reflects how in the digital environment several genres and modes can be integrated seamlessly to achieve several communicative purposes.

A frequent guideline for authors who want to publish video abstracts alongside a paper is that the video abstract should (preferably) be in English. Pajam Sobhani, the founder of *Latest Thinking*, justifies this requirement as a way to ensure maximum dissemination, considering the dominant role of English in academic communication (Enago Academy, 2018). He adds that the presenters of most *Latest Thinking* videos are 'non-native' English speakers, but asserts that they are not at a disadvantage because the videos are not produced live, so recordings can be redone. In fact, he considers video journals particularly useful for 'non-native' speakers, because they enable speakers who do not feel comfortable presenting at conferences to be physically visible.

However, while many presenters in video abstracts are speakers of English as an additional language (EAL), there is no evidence to suggest that this genre helps EAL researchers overcome language barriers. First, as an optional, supplemental promotional genre, it is likely that only authors who are confident in their ability to communicate in English will produce video abstracts. Secondly, to date no studies have compared the response (in terms of likes, or views, for instance) to abstracts produced by native speakers of English and speakers of EAL. It would, therefore, be interesting to explore whether this genre makes it easier for EAL researchers to share and promote their research. The possibility of producing bilingual videos, enabling researchers to present their papers in the language in which they are more proficient, could make this genre more attractive for EAL researchers. Although this possibility is not contemplated by *Latest Thinking*, the site is considering providing subtitles in other languages.

A genre related to video abstracts is that of AudioSlides. This genre is interesting because it sheds light on the emergence, evolution and disappearance of internet genres and on the fact that genres exist to respond to rhetorical exigences. AudioSlides was an innovation developed by Elsevier consisting of short webcast-style PowerPoint presentations about an article, displayed next to the article. Without peer review, this format enabled authors to use their own words to provide an overview of the article and make it available to a wide audience. For a period, the genre seemed to be unavailable, and the Elsevier website noted that authors would be invited to create AudioSlides when the feature could be available again. However, at the time of writing, the AudioSlides feature is no longer offered to authors on Elsevier's 'Author tools and resources' page. This situation raises the question of what makes online genres take off and flourish and why some may not succeed and thus disappear. Yang (2017) analyzed a corpus of AudioSlides in the natural sciences to identify the structure of these presentations and examined how authorial stance and engagement was expressed through personal pronouns. He found that, contrary to what the publisher advised, there was more emphasis on the knowledge gaps and the design of the research than on the findings, and that personal pronouns were used to project authorial identity rather than to engage the audience. Yang (2017: 41) suggests that if the purpose of the genre is to promote a paper, speakers should try to create rapport with the audience. However, more research involving journal readers' perceptions of the genre would be necessary to determine whether this genre met the audience's expectations.

9.2.2 The video methods article: Addressing the reproducibility crisis

A video journal is a publication of research-based work in a video format. Some journals, whose articles consist of or include a recording of the procedures followed during the research, aim to facilitate the reader's understanding and replication of the method. They thus help to address what has been called the 'reproducibility crisis' or 'replication crisis', that is, researchers' belief, supported by reports (Baker, 2016), that the experimental results of a large number of studies cannot be replicated because the description of the methods in the written article makes it difficult for researchers to grasp the necessary techniques to conduct the experiment (Henderson, 2013). In response, the video journal with the highest impact, the *Journal of Visual Experiments* (*JoVE*), was launched. Its founder, Moshe Pritsker, a researcher at Princeton University, was not able to reproduce an experiment involving embryonic stem cells that had been published in a biology journal by researchers in Edinburgh, Scotland. After flying to Edinburgh to watch a demonstration by the authors, Pritsker considered it necessary to use a more effective way of transferring this type of knowledge than text (Quinn, 2014). In the VMAs published in *JoVE*, the affordances of the audio and visual modes are harnessed to demonstrate the sometimes complex techniques that could help other researchers to reproduce the experiments.

The articles in this journal are composed of professionally produced videos[1] and accompanying text. As Engberg and Maier (2019: 185) put it, the videos are 'embedded in the hypermodal context of *JoVE*', that is, a context that combines multimodality (meaning is created through the combination of video, written text, colored boxes and pictures) and hypertextuality. The first element in the article is the video demonstrating experimental procedures. This is followed by a written text with the following parts: abstract, introduction, protocol, representative results (which usually include pictures or more videos), discussion, materials and references (Gross & Harmon, 2016). Although the parts appear to be similar to those in traditional printed papers, the protocol seems to be the most prominent part, consisting of a highly detailed step-by-step description, which complements the video. Hafner's (2018) analysis of this genre reveals how semiotic resources are orchestrated to achieve the purpose of the genre (i.e. to facilitate replication of the techniques and procedures) and how the video works together with other genres on the website as part of a genre system. The longest section of the video is the Demonstration, consisting of one obligatory move (Move D2, Describing steps) and three optional but common moves (D1 Introducing Apparatus/Materials, D3 Elaborating on steps and D4 Explaining/Evaluating steps). The purpose of this section is to demonstrate the procedures, key to the communicative purpose of the genre: to share techniques. This is achieved by combining the spoken narration with video of the researchers performing the procedures, and with text on the screen when additional information is needed.

Hafner's analysis provides interesting insights into VMAs, which may help to understand genre innovation and development involving online video genres for communicating knowledge. First, it highlights the hybridity of this genre which, making use of the semiotic resources available in the new medium, draws on several types of research genres, such as written methods articles, lab demonstrations and conference presentations. Secondly, it shows differences in audience engagement when comparing VMAs with traditional publishing genres and more specifically with the written RA: the researchers seek to interact with and engage the audience explicitly (e.g. wishing viewers well in their research), and engagement is realized through various semiotic resources, not only through language. What will probably lead to the perpetuation and further development of this genre is, as Hafner (2018) remarks, that unlike other types of visual genres (e.g. the GA), in the VMA the visual component, especially the moving images, is key to responding to a rhetorical exigence: to describe protocols accurately. This accurate description is effectively achieved not only through the orchestration of various semiotic elements in the video itself, but also through the interaction between the video and the other elements in its web context. For instance, while watching the videos, users are encouraged to engage with other materials on the web page (e.g. to access the software used in the experiment through a hyperlink, or to read the written protocol) (Engberg & Maier, 2019). Hafner (2018) refers to these multimodal interactions to explain the 'functional specialisation' (Kress *et al.*, 2001: 21) of different modes in the *JoVE* web page: the moving image provides a precise demonstration, the spoken narration provides reasoning and the written text provides a more complete account.

9.2.3 Online videos to communicate with a wider audience

Online videos can also be used as a tool to communicate with and engage wider audiences, including the lay public (Allgaier, 2018). This interest in attracting the audience has resulted in the publication of audiovisual genres for the communication of knowledge in a variety of platforms. Publications such as *Nature* or *Science* use videos to tell people about the research they publish in a visual way. These online videos can be embedded on several websites and shared on social media, which contributes to the widening of their audience. Videos are also used by several institutions and organizations to make knowledge available and to engage the public. For instance, the Charlie Foundation (which provides information about diet therapies for people with neurological disorders) hosts a series of video clips in its YouTube channel (Cavalieri, 2020), while Ocean Networks Canada (an initiative of the University of Victoria to study the ocean) uses live and archived video to engage the public with the ocean environment and direct viewers to other educational content on the website (Hoeberechts *et al.*, 2015).

Although most studies on researchers' use of video have focused on videos produced in English, non-Anglophone researchers also use their first language to produce videos intended to communicate science to the public (Luzón, 2019b), which is consistent with the use of the local language in outreach genres (McGrath, 2014). Allgaier (2018) draws attention to the case of the Russian website *ПостНаука* (*PostNauka*, meaning 'PostScience'; http://postnauka.ru/). This project began in 2012 with around 2000 videos by 2019, where researchers share scientific knowledge and ideas directly with the audience. In December 2013, a spin-off of *PostNauka* in English was founded under the name *Serious Science* (http://serious-science.org/) as a project whose website offers articles, interviews with scientists and videos. Interestingly, the founders of *Serious Science* state on its 'About' page that they developed the English spin-off to go international and create a global project, which suggests they perceive English to be a lingua franca. Participants in the project are researchers and scientists working in various fields who 'provide first-hand accounts of their fields of expertise', thus avoiding mediation and interpretation by journalists. The website states that the authors 'do their best to make their ideas sound, engaging and fully accessible' by speaking in 'plain and effective academic English' (http://serious-science.org/about).

Some research groups also produce videos in their local language, embedded in their websites or available on online platforms (e.g. YouTube, Vimeo), to explain important concepts in their disciplines. The use of the local language helps these groups to connect with the local audience and arouse public interest in their research activity and their discipline. For instance, a series of video clips called 'Nanotechnology capsules'/'Capsules of nanotechnology' is used by the Nanostructured Films and Particles (NFP) group (University of Zaragoza) to help students understand nanotechnology concepts, to disseminate knowledge in the field of nanotechnology and to promote scientific culture (Sebastian & Giménez, 2016). Although the videos have versions in both English and Spanish to reach a wider audience, the number of views is much higher for the Spanish than the English version. This example provides evidence of the important role of language choice in genres to communicate science to the lay public, and suggests the need to produce bi/multilingual videos to widen dissemination.

The increasing use of videos to communicate science to the general public makes it necessary to investigate how these videos can convey meaning effectively and achieve the purpose of engaging the audience. Several studies have analyzed different types of video genres – among them medical video clips from the YouTube channel of the Charlie Foundation (Cavalieri, 2020), or three-minute theses and author videos (Rowley-Jolivet & Carter-Thomas, 2019) – to identify strategies used by the producers to adapt the content to the new context and make it understandable and engaging for non-specialist audiences. These strategies include, for example, reformulation and repetition, questions to enhance interactivity and using a personal style in researchers' comments on their work. While most previous research has focused on the linguistic features of the videos, our study of two types of science videos (promotional videos produced by research groups and videos from the *Nature* Online Video Streaming Archive) (Luzón, 2019b) examines how linguistic resources are co-deployed with other semiotic resources (e.g. visual resources) to enact rhetorical intentions.

9.3 Research Case Study: An Analysis of Online Popular Videos to Attract the Local Audience

In this section, we present the results of a study of popular science web videos, that is, short videos focusing 'on the communication of scientific contents for a broad audience on the Internet' (Muñoz-Morcillo *et al.*, 2016: 1), produced by Spanish research groups. Public communication of science is becoming increasingly important not only for scholars but also for research groups, because of both the awareness that public interest contributes to the development of a discipline and the importance that funding agencies attach to the dissemination of results. Research groups are using the video format to disseminate information about their discipline and research by presenting it in ways that the general public can understand and potentially act on. In this study, we explore how researchers interact with the public in 15 online videos produced by research groups at the University of Zaragoza in collaboration with the Unit of Scientific Culture and Innovation (UCC+I) of the university. The UCC+Is at Spanish universities act as intermediaries between the institutions that host them and citizens and have become one of the main agents in the dissemination of academic knowledge and innovation in Spain. The UCC+I at the University of Zaragoza provides a repository of professionally produced videos made by researchers, which could be used as models for other institutions or organizations involved in knowledge dissemination. The videos are the outcome of the 'Workshop on scriptwriting and production of scientific documentaries', an initiative to help researchers produce short science documentaries which can be used in several contexts (public screenings, conferences, teaching, dissemination in high schools) and which are hosted on the websites of the UCC+I (https://ucc.unizar.es/taller-de-guion-y-produccion-del-documental-cientifico) and of the research groups. The videos aim to make the work of the groups more visible to the local audience and to promote scientific, technological and innovation culture in society, which explains the choice of language, Spanish.

To determine how various semiotic resources are combined by these research groups to recontextualize scientific discourse into digital videos and to engage the lay public, we sought to answer the following questions:

(1) What is the rhetorical structure of these videos and how are their arguments organized?
(2) How are various semiotic modes combined to make the content understandable?
(3) How is the credibility and authority of the research group constructed?
(4) What devices are used to engage the reader and to evaluate the research?

To answer these questions, we drew on Hyland's (2010: 117) concept of 'proximity', i.e. 'how writers represent not only themselves and their readers, but also their material, in ways which are most likely to meet their readers' expectations'. However, as online videos combine several semiotic modes to convey meaning, we extend this model to include not only text but also other semiotic devices which enable the producers to signal their attitude towards the content and establish a relationship with the viewer (see also De Groot *et al.*, 2016; Hafner, 2018; Luzón, 2019b).

(1) Rhetorical structure of the videos

All the videos analyzed consist of documentaries where storytelling is crucial. For instance, the video *The Turtle of the Desert*[2,3] tells the story of research carried out to explain the presence of water turtles in a desert area. Such narrative films have the advantage for science communication of enhancing the persuasiveness of the message. They can 'visually transport people to places and situations they might otherwise never experience' (Norman, 2000: 28) and thus make the viewer care about scientific issues and take action. The videos were constructed in two ways: a commentator (or voiceover) introduces the members of the group (who explain aspects of their research); or, more frequently, a member of the group tells the story and introduces the other members. All the videos have a similar rhetorical structure, responding to their purpose of engaging the general public:

(1) A brief introduction to attract the viewer's interest, usually consisting of an eye-catching scene involving a question to be answered, a mystery to solve or a narrative story.
(2) The title of the video.
(3) The presentation of the group and their work, involving the explanation of concepts, methods and techniques.
(4) A conclusion, which usually goes back to the beginning, repeating images or linguistic references to the beginning of the video, and/or discussing the relevance and importance of the research and/or presenting the applications of the research in everyday life.
(5) Closing credits.

(2) Strategies to make the content understandable

These videos orchestrate visual and verbal semiotic resources to make the message more understandable for the lay audience. As in other types of popularizations, the producers make connections to what viewers are likely to know, and use non-technical language to define or explain unfamiliar concepts or methods (e.g. 'squalene', 'zeolites'), or to clarify what is meant by a term that the viewers already know (e.g. what 'extra virgin oil' really means). These linguistic resources are often co-deployed with visuals of different types: images of the object being defined, animations to explain a process, drawings to illustrate explanations, graphics and maps.

(3) Strategies to construct the credibility and authority of the research group

Several strategies bestow credibility on the groups, some shared with those used in popular science texts. In popularizations, the credibility and authority of scientists are constructed by describing their position in an institution or by using quotations to allow researchers to talk about their research (Hyland, 2010). Similarly, a strategy used in videos to establish the credibility of the researchers is to draw attention to their status and affiliation by using various modes, for example by identifying the position of the researcher telling the story (through superimposed text when the researchers first appear on screen); by explicitly mentioning the prestige of the researchers, the research group or the research center; or through photographs of the faculty or research center.

The Presentation section usually begins by identifying the group, to claim authority over the research (e.g. 'The research group of plant-based foods from the University of Zaragoza has been conducting studies on foods as varied as …'). Credibility is also constructed by representing the researchers as experts through the interplay of visual and verbal resources. In all these videos, the researchers themselves present their research and display their knowledge of the discipline. In one of the videos, *Liquid Crystal*,[4] the group member who acts as narrator asks another member to explain what a liquid crystal is (see Figure 9.2). The researcher explaining the concept wears a lab coat and uses the blackboard in a classroom to provide visual support for her explanation, which helps viewers perceive her as an expert. The videos represent researchers not only explaining science, but also 'doing' science. Live-action footage represents members of the group 'doing research' in their research setting and using the methods of their discipline, for example, digging in archaeological sites, or using laboratory equipment or computers. Like the VMAs analyzed by Hafner (2018), these videos, which show different members of the group engaged in research, emphasize the distributed expertise of research groups, in contrast to other conventional academic genres, like the RA, where the agency of individual authors is not visible.

Figure 9.2 Screenshot from *Liquid Crystal* (https://ucc.unizar.es/enziende-la-ciencia/17346) (Universidad de Zaragoza) (Licensed under Creative Commons Attribution 4.0 International)

Figure 9.3 Screenshot from *Virtual Archeology* (https://ucc.unizar.es/enziende-la-ciencia/4826) (Universidad de Zaragoza) (Licensed under Creative Commons Attribution 4.0 International)

The use of exclusive 'we' (i.e. use of the pronoun 'we' to refer only to the researchers) also helps to project the voice and authority of the group. In these videos, exclusive 'we'-pronouns perform mostly the discourse function of 'recounter' or narrator of the various steps of the research process. This verbal recounting, usually juxtaposed with images of group members carrying out research, shows that the researchers are familiar with disciplinary procedures and methods, thus helping to construct their identity as experts. For instance, in the video *Virtual Archeology*,[5] one of the researchers explains how the group creates a virtual reconstruction of a Roman city by combining 'we' (e.g. 'We start from an idea, which the archaeologist has; we generate the first sketches') and computer simulations representing each step (Figure 9.3).

(4) Devices used to engage the reader

The videos we analyzed in this study display a high frequency and variety of engagement devices (see Hyland, 2005) used by the producers to connect with the audience and enable them to participate in the text: appeals to shared knowledge and experiences, personification, narrative style, informality, inclusive vocabulary and inclusive verbal forms, reader mentions, questions and directives. These engagement devices are realized through various semiotic resources.

Engagement markers are particularly relevant in the introductions to the videos. In this section, several linguistic, visual and auditory resources are used to create an atmosphere familiar to the general public and to raise issues of concern to this public. These resources include the use of familiar sounds and images to begin the video (e.g. the sound of birds chirping or of water flowing in nature). The video *The Journey*

Figure 9.4 Screenshot from *The Journey of the Steak* (https://ucc.unizar.es/enziende-la-ciencia/17345) (Universidad de Zaragoza) (Licensed under Creative Commons Attribution 4.0 International)

of the Steak,[6] where the producers explain the measures being taken to make meat safe and avoid the outbreak of diseases such as 'bovine spongiform encephalopathy' (mad cow disease), begins with the sound of potatoes being fried followed by a film of a man in a kitchen cooking a steak for a young girl, who is sitting at the table waiting to be served. A voice on the radio mentions mad cow disease, which makes the man wonder about the safety of the steak (a concern viewers may share) (see Figure 9.4).

Another frequent strategy is personification. By focusing on a particular individual, the viewers ideally identify with them and feel empathy. For example, in the video *A Journey to Immunity*,[7] the audience is 'represented' by a member of the lay public, who asks questions that are relevant for this type of audience. Most of these videos are constructed to make viewers identify with the characters and feel empathy with their interests and goals, a strategy that, as Finkler and León (2019) note, helps to increase narrative persuasion. In all the videos, the short physical distance between the researcher and the character representing the audience creates proximity (see Figure 9.2), which helps to convey a feeling of intimacy and bridge the social distance between experts and lay audience.

The narrative style of many of the videos, in which viewers are told the 'story' of the research, is also used as a strategy to create interest and engage the lay public. For instance, the video *The Turtle of the Desert* begins like a story, where the research is depicted as the steps followed to solve a mystery (e.g. Example 9.1a). Similarly, the researcher acting as narrator in the video *The Ricla Crocodile's Curse*[8] states that he is going to tell a story (e.g. Example 9.1b).

Example 9.1

(a) June 1989. A naturalist once again begins his journey through the desert environments of the Bárdenas Reales (…) He discovers something out of the ordinary [*the camera focuses on a turtle*]. This turtle forgotten for 17 years is the beginning of our story.
(b) This is the story of a curse and a group of researchers who fought against it.

Informality, that is, the use of distinctive features that express a personal tenor and create a closer relation with the audience, such as first- and second-person pronouns to refer, respectively, to the speaker and the addressee (Hyland & Jiang, 2017: 42) (see Chapter 3), is also used to bring the researchers close to the public and therefore engage them. Researchers are referred to by their first names and are pictured in casual clothes and using informal language. For instance, in the video *The Ricla Crocodile's Curse*, where geologists (the Aragosaurus group) explain the difficult process that they have followed to identify a crocodile fossil, a young female researcher, dressed in a T-shirt, talks about the relevance of the finding. She uses colloquial language with features typical of conversation ('What is cool, well, the first thing is that it is a unique genus and species'). Questions are also used to create interest and empathy on the part of the viewers, particularly at the beginning of the video. Most videos have an explanatory structure, consisting of a starting question, usually highly relevant or interesting for the general public, and an account of how the research carried out by the group helps to answer that question. The video *Coexisting with Water*,[9] for instance, begins with the questions: 'Can we understand and predict the behavior of water? Can the effects of floods be avoided or mitigated? Do we know how to coexist with water?'. And at the beginning of the video *The Journey of the Steak*, the viewers hear a man wondering, 'Will this steak be safe?' before giving it to his daughter.

Another frequent strategy to engage the audience is the use of inclusive vocabulary and inclusive verbal forms (Example 9.2). These elements contribute to establishing affinity and intimacy with the viewers. As also used in TED Talks (Scotto di Carlo, 2014), they help to bridge the gap between the researchers and the audience by transmitting the message that they share common interests, and to represent the viewers as participants in research.

Example 9.2

(a) At this moment ... we can predict in real time what will happen when heavy flooding approaches and we can minimize the damage that will occur. (*Coexisting with Water*)

(b) As children, we dream of being soccer players, heroes, etc. and at some point in our life we discover our vocation. (*Teledetection: Discovering Invisible Territory*)

Researchers in these videos also use direct appeals, that is, questions directly addressed to the audience (e.g. 'Have you ever wondered what makes these screens possible?') or directives (e.g. 'Imagine'). In many cases, the directives are addressed to a member of the research group representing the audience to attract attention to a particular visual (e.g. 'look'), and the audience sees a close-up shot of what this member is seeing. In some videos more indirect directives are also used in order to appeal to the general public to take action, as illustrated in Example 9.3 from the video *The Turtle of the Desert.*

Example 9.3 The conservation of species and the natural environment should not be the work of a few. Researchers, administration, educators, local population ... We must look for common lines to improve the planet on which we live.

(5) Strategies to express attitude and commitment to the propositional content

To make the research valuable for the audience, videos draw on the 'application appeal', that is, presenting the research as having further benefits or future applications (Fahnestock, 1986), which is realized by combining verbal and visual semiotic resources. There is an abundant use of evaluative words to emphasize the importance and relevance of the group's research, focusing on its usefulness and benefits for the general public (e.g. 'Research on transmissible spongiform diseases is very important because these are diseases that are lethal' in *The Journey of the Steak*). The verbal presentation of uses or applications is synchronized with video footage representing these applications, which helps the viewers visualize them.

These videos follow principles that, according to Finkler and León (2019), are key for communicating science effectively in what they call 'SciCommercial Videos'. Drawing on marketing communication techniques, Finkler and León (2019) developed a conceptual framework for storytelling through videos about science. They propose that memorable successful science ideas presented through short-format videos have the following features: simplicity (i.e. prioritization and focus on the core idea, rather than on scientific details); unexpectedness and surprise, which can be achieved beginning with a question; concreteness (i.e. concrete words that evoke distinct images); credibility of the source and likeability of the characters, which lead to empathy; authentic reactions and emotions; hooks that make people connect with science (e.g. stories, affective imagery); and storytelling that leads to action.

The videos analyzed here were designed to promote scientific culture among the general public; however, the groups also seek to promote their discipline in order to attract students (the videos are shown in secondary schools) and to promote their research to get funding (dissemination activities are important to obtain national and local funding in Spain). These imagined audiences and purposes account for the important role of engagement in these videos, revealing a shift from monologue (still frequent in many online videos to communicate science) to viewers' involvement. The combination of modes enabled by the video format affords new ways to make meaning and to engage the audience. Visuals in these videos are not included in order to explain difficult concepts and methods (as in other genres, like the VMA) or to provide evidence for verbal arguments, but rather to facilitate understanding of the verbal account and engage the reader. The verbal narration is integrated with images which help viewers visualize the stages of the narrative and attract their attention.

9.4 Concluding Remarks

In this chapter we have explored the relationship among genre, mode and medium when communicating knowledge in the digital environment, and provided an overview of digital genres for knowledge communication that harness multimodal affordances, focusing particularly on audiovisual genre innovation. These genres enable researchers to respond to new rhetorical exigences in their communication with peers (e.g. reproducibility of research, sharing complex techniques) and to engage the interested public and promote science literacy. The digital environment makes it possible to combine various modes and connect different media on a single platform to facilitate communication of meaning. It also makes it possible for researchers to embed their (audio)visual material on various websites and to share it on social media, thus increasing its visibility and outreach. We have also seen that, although in videos that target international research communities English seems to be the default language, when composing videos to promote science literacy and their research locally, some researchers resort to their local language in order to engage the local audience. An underexploited affordance of the digital medium is the possibility of composing multilingual videos.

The emergence of the genres discussed in this chapter offers new possibilities for knowledge communication, but the multimodal composing that they involve also poses new challenges for scholars, an issue that requires attention in further research. In order to communicate knowledge effectively using these genres, scholars need new literacies to combine multiple modes for meaning making and to compose multimodal texts for a variety of purposes and audiences. In addition, the use of online videos for research visibility and promotion also raises equity issues, since many scholars in low-income countries may lack the technical resources needed to produce them themselves or the funding to pay the services for professionally produced videos.

Notes

(1) The high cost of video production and the technological infrastructure required is a serious limitation that restricts access to this genre. *JoVE* is a subscription journal (although it was originally open access) and researchers have to pay a fee to publish, although the journal may offer fee waivers for researchers whose funding comes from institutions in low- or middle-income countries.
(2) The titles of the videos and the examples have been translated into English.
(3) https://ucc.unizar.es/enziende-la-ciencia/17344.
(4) https://ucc.unizar.es/enziende-la-ciencia/17346.
(5) https://ucc.unizar.es/enziende-la-ciencia/4826.
(6) https://ucc.unizar.es/enziende-la-ciencia/17345.
(7) https://ucc.unizar.es/enziende-la-ciencia/12634.
(8) https://ucc.unizar.es/enziende-la-ciencia/4818.
(9) https://ucc.unizar.es/enziende-la-ciencia/17347.

10 Assessing Research and Participating in Research Discussions Online

10.1 Introduction

The imperative of open science and the possibilities enabled by technological advances have resulted in a radical transformation not only of how knowledge is produced, communicated and commercialized, but also of how it is evaluated by peers and legitimized. Web 2.0 technologies and their affordances for openness and interaction offer myriad possibilities for the discussion, debate and evaluation of scientific ideas and research results by peers both before and after publication (Sugimoto *et al.*, 2017; Tattersall, 2015). Several digital genres and platforms (e.g. Twitter, blogs) facilitate the crowdsourcing of peers' opinions and reviews of work in progress and pre-prints on online servers (e.g. ArXiv and BioRxiv). The internet also offers tools for open peer review and comments provided on publisher-based platforms or journals. Digital genres also play an important role in the process of validating or refuting published results. Although there have always been genres intended for post-publication evaluation (e.g. letters to the journal editor, book reviews, commentary articles), digital genres such as blogs or tweets help to accelerate the process of evaluating published research and make it possible for other types of audiences, not only experts, to discuss and evaluate research findings (Buehl, 2016; Luzón, 2014; Sidler, 2016).

However, even if the internet facilitates a more participatory and transparent evaluation process, one not limited to reviewers selected by the editor, this same openness may also involve risks and challenges for authors (e.g. to become a target for trolling or the tarnishing of their identity and reputation if their work is criticized in an open forum) and for reviewers (e.g. their reputation is also at stake) (Brunner, 2016; Taylor & Francis Group, 2015). The openness and ease of commenting and responding afforded by the internet may have negative unintended consequences for knowledge construction and validation, as some comments by non-expert audiences may be just ill-informed opinions or tangential comments, and they can help to keep a refuted idea in circulation and thus contribute to confusing 'the state of knowledge in a field' (Fahnestock, 2016: 137).

In this chapter, we first discuss digital genres used for online peer review of manuscripts submitted for publication, and digital genres for post-publication evaluation and their potential impact, both positive and negative, on quality assessment processes. Then, we examine the role of blogs and social media as sites used to raise concerns about the validity or honesty of published research or to engage in scientific controversies. Finally, we present the results of a study which analyzes the role of academic blogs in engaging in scientific controversies and the language used by participants to evaluate others' claims and put forward their position in such controversies.

10.2 Open Review: Emerging Genres for Open Review and Post-Publication Evaluation

Peer review has a central role in both controlling the quality of research and certifying it as scientific knowledge and gatekeeping the formal publication of results (Lillis & Curry, 2015; Paltridge, 2017; Puschmann, 2015a). In the last few years, the traditional model of blind peer review has been questioned on several ground, such as lack of transparency, which may lead to unfair reviews. The internet offers the possibility to transform the process, aligning it with the principles of open science, that is, making it more transparent and reliable, and increasing its accountability, although concerns have also been raised about bias against certain author demographics when the identity of the authors is known to reviewers (Tennant *et al.*, 2017).

The term 'open peer review' has been used to refer to different concepts, including open authors' and reviewers' identities, open review reports (i.e. published with the article), open participation of the wider community in the review process, interaction between author(s) and reviewers, and/or between reviewers, making manuscripts available (on pre-print servers) before formal peer review, and open review or commenting on publications (Ross-Hellauer, 2017: 5). These forms of open peer review have prompted the emergence of new review genres and influenced the features of migrated genres.

10.2.1 Genres for open peer review before publication of final version

Several journals and online platforms are publishing reviewers' reports (with the identity of the reviewer disclosed) and authors' responses along with the accepted papers (and sometimes their previous versions). This is the case of the *BMJ Open* (Görögh *et al.*, 2019), which includes links to this information within the publication history, or the medical journals in the BMC series of BioMedCentral. All EMBO Press journals publish an optional 'peer review process file' for each article, which includes the anonymous reviewers' reports, the authors' responses and the editorial decision letter (*EMBO Journal*, 2020). The platform *F1000Research* has also opted for open identities and open reports. Articles that pass an in-house check to ensure quality (e.g. originality, readability and author's eligibility) are accepted, published immediately and reviewed openly by expert named reviewers. Authors can then respond openly to these reports and revise their paper (*F1000Research*, 2020). Although the purpose of these open reports is the same as that of the traditional referee report, considered by Swales (2004) as an 'occluded genre' (i.e. 'out of sight' to outsiders), the fact that they are open for everybody to evaluate and that in many cases the identity of the author is disclosed may have an effect on the features of the genre, as we discuss below.

The internet also facilitates open interaction among different participants in the process, which may also lead to genre innovation and change. For instance, *Frontiers* (2020) established a 'collaborative review forum', where editors, reviewers and authors can interact in real time. Similarly, *eLife* makes use of the 'online consultation session' (see Schekman *et al.*, 2013), a format that could be adopted by other journals that engage in the practice of open peer review interaction. During this session, the editor and the reviewers discuss the manuscript before reaching a decision (Görögh *et al.*, 2019). As stated in the journal's website (*eLife*, 2020), the aim of the online consultation session is 'to provide clear and decisive instructions to authors, so that they know what they need to do to get the article published'. If the author is asked to revise the manuscript, the decision letter includes one set of instructions, but the author does not receive the full reviews; if the manuscript is rejected, the decision letter includes the full reviews, which explain the reasons for rejection. As inferred from this description, the adoption of this practice by other journals might also affect the form of the traditional editorial decision letter (and peer review reports), which might evolve into two different genres, depending on whether the paper needs to be revised or is rejected, with two different exigences: a genre intended to provide clear, non-contradictory *instructions* for revision; and a genre intended to provide *justification* for the editorial decision to reject. This could improve the efficiency of the peer review process, as the current practice of providing authors with the reviews regardless of the decision does not always send a clear message to them because the reviews might include contradictory comments or mixed messages about publishability.

Some publishers are fully using online review genres and combining them in the evaluation process. A prominent example is the European Geosciences Union (EGU) and its journal *Atmospheric Chemistry and Physics*, which follows what Pöschl (2012) calls a 'multi-stage open peer review'. Manuscripts that pass a rapid in-house quality check are made available immediately as 'discussion papers' in the journal's discussion forum, to be reviewed and commented on by selected referees and members of the scientific community. These comments and the authors' replies are published online with the paper. A second stage is a traditional peer review, after which the paper, if accepted, is published in the main journal. This interactive peer review and public discussion may again be the origin of a new review genre, which, as pointed out by Pöschl (2012: 140), may have advantages for all parties: authors are offered direct feedback while plagiarism is reduced; referees and readers are provided with documentation of critical comments, controversies and information that complements the research article, which may be particularly useful for non-specialist readers; and publishers minimize the waste of time because areas for improvement are identified before the manuscript passes to the stage of formal peer review.

Despite the potential benefits of the open review practices described here, they may involve risks and challenges for scholars, particular for junior or historically marginalized researchers. As reviewers, such scholars may feel more reluctant to honestly evaluate the work of more established researchers who could influence their career in the future (e.g. members of editorial boards); as authors, they may refrain from submitting papers because of the fear of receiving negative reviews in a public forum (Taylor & Francis Group, 2015; Tennant *et al.*, 2017); in addition, providing the reviewers with information about the authors may also lead to gender bias or bias against multilingual scholars (Budden *et al.*, 2008; Ross *et al.*, 2006). Furthermore, if open peer review is conducted in journals behind paywalls, the possibility to participate is limited to researchers who have access to a subscription. All of these concerns raise the issue of equity and emphasize the need for more research on whether open peer review facilitates or impedes marginalized scholars' participation in global academia.

In addition to submitting manuscripts to journals for open peer review, researchers in certain disciplines can make pre-peer-review versions of their research articles openly accessible on subject-specific pre-print servers (e.g. ArXiv or BioRxiv) or general purpose repositories (e.g. Zenodo, Figshare), which facilitates informal review and discussion before formal publication. One of the benefits of posting manuscripts on pre-print servers is that the discussion of ideas in this phase between authors and readers can help authors make improvements before the publication of the accepted final version of the paper, which can accelerate knowledge distribution (Bernstein, 2015). Findings can be evaluated by peers quickly and openly, which is particularly important at moments when the availability of valid findings is essential to make rapid progress in finding solutions. A clear example is the pre-prints being shared on these servers to help contain the spread of Covid-19. While these preprints are not peer-reviewed and therefore some of the findings may be flawed, they are disseminated quickly through several outlets (Kelland, 2020), which makes open review by other researchers particularly important. To draw awareness to the unvetted nature of these manuscripts, BioRxiv added a warning at the top of pre-prints reporting on coronavirus research that 'these are preliminary reports that have not been peer-reviewed' and should not be considered or reported in the media as established information.

This warning was published after complaints about the publication of a pre-print in BioRxiv which, according to some commenters, seemed to suggest that the coronavirus was related to HIV, the virus that causes AIDS, and had been designed by humans. A month after the pre-print was posted on BioRxiv, it had 120 comments, many of which challenged its conclusions by stating that the methodology was flawed and careless and claiming that the results were faked or merely coincidental. Although most of the comments were anonymous, and therefore their authoritativeness could be questioned, some commenters provided their academic credentials to bestow credibility. Some of the comments were criticisms of the claims made in the pre-print that were supported by laboratory work such as replication of the experiment (e.g. 'I blasted each of the four 2019-nCoV inserts shown in Table 1 and received 100% identity with a number of other hits other than HIV-1') or by long arguments based on disciplinary knowledge. The authors of the pre-print used the comment feature to respond to criticism, explain the problems with the paper as a consequence of the need to publish Covid-19 research results quickly, defend their honesty and announce the withdrawal of the paper and their intention to revise it taking the comments into account. After the authors' comment, a retraction note appeared on the BioRxiv website. As Oransky and Marcus (2020a) remark, the reaction from the scientific community to the claims in the paper and the consequent retraction were swift, which reveals how the peer comments in pre-print servers can make corrections in science faster and more effectively than can traditional peer review. Additionally, the comment by the authors suggests that this process of participatory review may contribute to improving the paper by allowing the authors to get feedback by a high number of peers.

10.2.2 Post-publication open review

In addition to the evaluation of and commentary on manuscripts (including pre-prints), the internet also facilitates open review and critique after the formal publication of peer-reviewed papers, with the emergence of new genres that complement traditional genres for feedback on published work, such as letters to the editor or commentary articles. For instance, *Faculty Opinions* is a database that offers post-publication recommendations of worthwhile articles in biology and medicine. A group of more than 8,000 'faculty members' (selected experts) recommend and review important articles in their subfields (Akers, 2018; Sugimoto *et al.*, 2017).

The platform is not intended to refute claims, but rather is publicized as a service that enables researchers to find articles recommended in their fields and also to get an 'easily digestible summary of why an article has been recommended' (*Faculty Opinions*, 2020). The benefits for scientists, according to the platform, is that through this service they can 'discover the articles of most relevance to [their] research, evaluate expert opinion on the most impactful research, [and] understand the key points and context of the article' (*Faculty Opinions*, 2020). The review includes a star-rating recommendation, one or more tags for the article (e.g. *new finding*, *interesting hypothesis*), and a positive review of the article, of variable length (although typically brief) focusing on the reason for its recommendation. As with most genres of online evaluation, these reviews are not stand-alone documents but part of an ongoing evaluation system: recommendations are cumulative, as articles can get more than one recommendation (as Figure 10.1 shows), and a recommendation can be responded to by a dissenting opinion. The reviews also include the 'Faculty Opinions Score' (65.1 for the article in Figure 10.1), a new metric intended to provide a more robust measure of the quality of the article, which is obtained by combining the expert recommendations of the article with bibliometric data for that article.

The recommendation reports on this platform, therefore, respond to rhetorical needs that are different from those met by the traditional review report, whose filtering function predominates over its formative function. The *Faculty Opinions* report can be regarded as a new genre of open non-anonymous peer review and recommendation/endorsement, similar to other conventional genres such as the book review. This genre is related to other tools on academic social networking sites (ASNSs) and online journals which provide alternative impact metrics of publications (mentions, number of reads/downloads, likes, recommendations) (Pöschl, 2012), as demanded by the measurement school of open science (see Chapter 3).

Other tools for open post-publication review are 'reader comments', which can be offered by journals (e.g. *PloS One*) or by independent peer review platforms (e.g. *PubPeer*). In the journal *PLoS One*, for instance, readers can discuss the content, conclusions and significance of published papers by using 'reader comments' (Casper, 2009, 2016; Kelly & Kittle Autry, 2013). Although opening the commenting process to the public may have benefits, it also has drawbacks such as the possibility of anonymous comments. On the peer review platform *PubPeer*, published authors (who have to be verified as such) can comment anonymously on published papers (Brunner, 2016; Townsend, 2013), thus raising concerns related to their objectivity and transparency (Brunner, 2016; Torny, 2018). The anonymous commenting enabled in this platform, which seems to be at odds with the principles of open science, may lead to ungrounded intense criticism of some papers. In addition, although

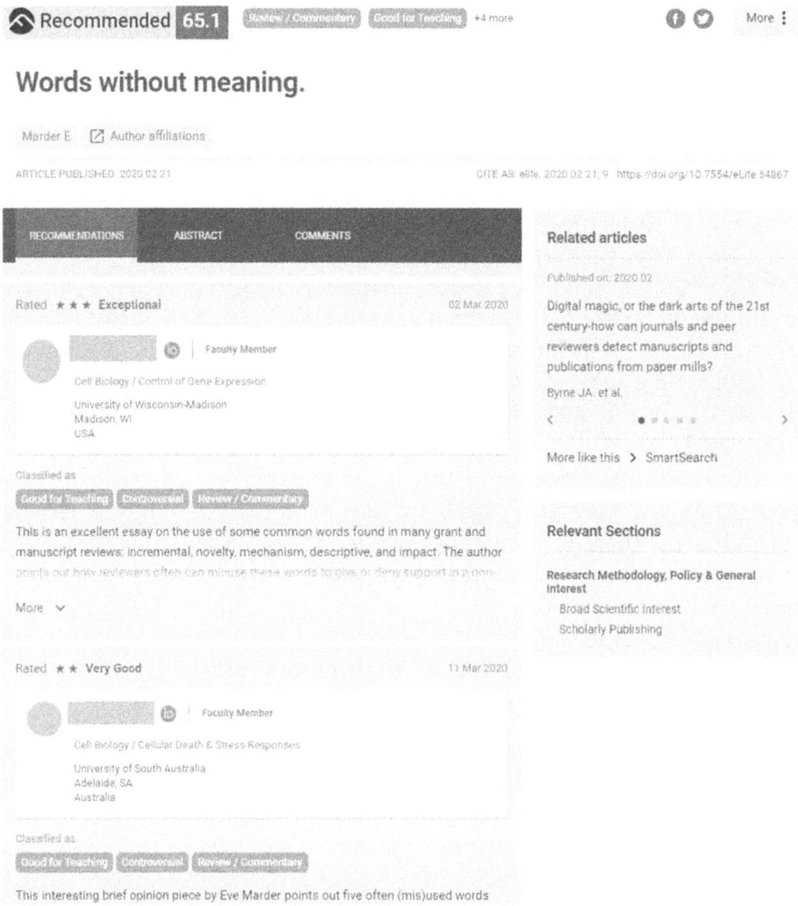

Figure 10.1 Screenshot of recommendations for the paper 'Words without Meaning' (https://facultyopinions.com/prime/737407597).

these comments may have resulted in the exposure of high levels of research misconduct (since commentators are protected by anonymity), as claimed on the 'About' page of this platform, such exposure may also have serious consequences for researchers' careers.

All these tools for post-publication review promote a conception of open review as an 'ongoing process' for the improvement of research output rather than a 'distinct process that leads to publication' (Ross-Hellauer, 2017: 12). This may, in turn, lead to changes in the genre of the article and in writing for publication practices. An interesting example of these innovations taking place is the Living Review genre published in three Springer journals (Ross-Hellauer, 2017), and recently adopted by other journals including the *British Medical Journal*. They consist of review articles that the authors keep up to date by incorporating new

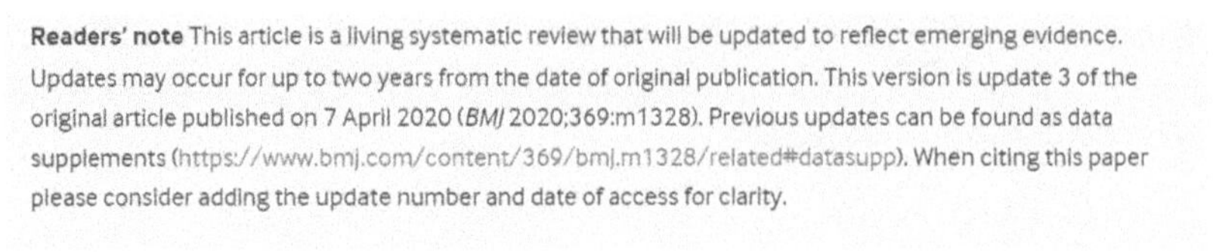
Readers' note This article is a living systematic review that will be updated to reflect emerging evidence. Updates may occur for up to two years from the date of original publication. This version is update 3 of the original article published on 7 April 2020 (*BMJ* 2020;369:m1328). Previous updates can be found as data supplements (https://www.bmj.com/content/369/bmj.m1328/related#datasupp). When citing this paper please consider adding the update number and date of access for clarity.

Figure 10.2 Note in a living review in *BMJ* (https://doi.org/10.1136/bmj.m1328) by Wynants *et al.* (2020) (Licensed under Creative Commons Attribution [CC BY 4.0])

developments in the field. In Springer journals, the article history on the article's web page includes a link to the original version and a summary of changes (e.g. New Sects. 3.8 and 3.9; extended Sect. 6. Added new Figs. 8, 10). In the *BMJ*, links are provided to the previous versions (see Figure 10.2).

The rhetorical exigence of this genre is explained by the editors of the *BMJ* (Macdonald *et al.*, 2020: m2925) by arguing that in fast-moving research areas there is 'an urgent need for timely access to high quality, up-to-date syntheses' which aggregate and evaluate research that examines a particular question.

10.2.3 The effects of web-based affordances and openness on the features of review genres

Research on open review has focused on the different types and systems of implementation. However, little attention has been paid to date to how openness and web-based affordances might affect genre innovation and change and transform the ecology of genres for knowledge production, communication and evaluation. The status of the research article and its format may undergo change in response to the possibility of public discussion of the accuracy and quality of the reported research. As Gross and Harmon (2016: 138) insightfully point out, unlike traditional peer review, open peer review 'preserves the true knowledge status of the paper in flux' because debates about the quality and significance of the paper are publicly visible and are susceptible to be reopened at any moment. Casper (2009, 2016) and Pöschl (2012) also suggest that the discussion through commentaries in interactive peer reviews may lead authors to write articles differently, that is, to present arguments more persuasively in order to avert potential criticism and public loss of face. Pöschl (2012: 140) contends that authors may make extra efforts to improve the quality of their manuscripts as a result of the awareness that errors or weaknesses in experiments or interpretations are more likely to be noticed during open interactive peer review.

The effects of web-based affordances on review genres was studied by Casper (2009), who examined four online tools for post-publication feedback and review: e-letters and traditional letters to the editor (published also online) in *Science*, both of which include names and affiliation and are vetted by the journal's editor; and comments and notes in the open journal *PLoS One*, which are often pseudonymous and are only vetted after publication if concerns about incivility arise. Casper (2009) found that comments and notes are shorter and more informal and dialogic, and present a wider range of speech acts (e.g. corrections, suggestions, questions) than the e-letters and letters to the editor in *Science*, where politeness conventions and negotiative rhetorical tactics are more prominent. As a new review genre, notes are intended to provide feedback and critique on specific aspects of published papers. Their emergence has been possible thanks to the annotation standards for digital documents and to tools that enable sentence-level evaluation by peer readers. These affordances help to explain the specific features of the genre. Notes do not need to establish context because they are attached directly to a specific section of the text on which they comment. Casper (2009: 88) compares them to comments/questions in a lab or seminar, used to provide more informal and shorter feedback than online comments or to request clarification of particular aspects (e.g. the note 'Should this be molasses? Apologies if I'm wrong', attached to a passage in the methodology of an article). However, the affordances of the internet confer these notes with permanent status (unlike spoken comments). Notes can also be posted by article authors to add brief information, such as links to supplementary material or corrections of typographical errors or in response to previous notes or commentaries. They are therefore a versatile tool.

The small body of research comparing traditional closed review reports and open peer reviews has found differences in terms of the nature of interaction and the style of reviews that can be partly attributed to the public nature of open reviews. A remarkable difference noted by Bornmann *et al.* (2012), after comparing closed and open peer reviews in two chemistry journals, is that open peer reviews (both anonymous and attributed) were more likely to result in constructive interactions among the parties (author, reviewers, the wider community) and thus be more likely to improve the paper than were closed reviews. Open peer reports were found to be more cognitively elaborated than traditional closed peer review reports, thus longer. Their style was also more tentative and civil, with a higher frequency of hedges and a lower number of emotional words, and interlocutors are often addressed directly, simulating face-to-face conversation. Bornmann *et al.* (2012) also suggest that the added elaboration and the tendency toward civility may result from the openness of reviews to the public themselves, which means they can be evaluated.

Research on evaluative language in reviewers' reports suggests a relation between open reviewer identity and the use of politeness strategies. For example, in her analysis of peer reviews received by Slovak biomedical researchers writing in English, Kourilová (1998) found that politeness conventions appear to be different from those in most other genres of academic writing, with a high number of examples of blunt criticism, and fewer examples of mitigated criticism. Kourilová attributes this finding both to the power difference between author and reviewers and to the fact that the anonymity of peer review encourages the use of a more direct and confrontational style, which would be face-threatening if used in a public forum. This interpretation is supported by Nobarany and Booth's (2015) analysis of politeness strategies by authors of open-peer reviews. They found the frequent use of politeness strategies to mitigate criticisms, which seems to indicate that making criticisms in signed open reviews may require more politeness work to soften face threats than in closed reviews. The authors also suggest that in an open process the research community acts as a watchdog to control incivility resulting from unequal power relations. In line with previous results, Breeze's (2019) comparison of confidential author responses to referees and responses in an anonymized online open access review system documents that open author responses also involve more relational work and are thus longer and more elaborated. Interestingly, the open peer review in Breeze's analysis did not lead authors to show more compliance with referees' wishes. To the contrary, authors tended to express disagreement more frequently than in confidential peer reviews. Breeze (2019) suggests that when criticized in the public arena, authors may feel the need to defend themselves and show their competence as researchers.

The question of whether the features of these new genres for online feedback have a limiting or facilitating effect on peripheral scholars' participation in global scholarship has not been well researched. Research has revealed reviewers' bias towards peripheral scholars when these seek publication in English-medium journal, which may result in harsh feedback (Lillis & Curry, 2015; Walbot, 2009) or in hypercorrection, or 'intolerance of what are considered to be non-standard (…) forms' (Lillis & Curry, 2010: 164). Breeze (2019) surmises that the transparency afforded by open reviewers' reports published online might discourage reviewers from being influenced by factors such as nationality when recommending papers to be published, since they would need to justify their decisions better. However, there is little research on whether reviewers are more careful not to use insensitive language or indulge in hypercorrection when writing open reports or comments on papers by multilingual scholars.

10.3 Blogs and Social Media to Evaluate Published Research

As evidenced by the example of the retracted paper discussed in Section 10.2.1, social media may have an important role in post-publication evaluation. In addition to the various genres that are emerging to facilitate post-publication open peer review, readers of articles can publish their comments and thoughts in a variety of forums on the internet (e.g. blogs, Twitter, ResearchGate).

As with other platforms for peer review discussed above, web affordances make blogs and social media highly suitable outlets for new forms of evaluation and feedback in academia. Their main advantages, when compared with traditional peer review, is that they facilitate interaction and open participation in the review. Twitter, particularly, allows rapid and widespread dissemination of findings and real-time low-effort interaction and critique, which is probably a reason it has become popular among researchers (Mandavilli, 2011; Yeo *et al.*, 2017). In addition, the use of links, embedded photos or videos and hashtags enable authors to contextualize their blogpost or tweet or provide further information to interpret it and support the author's critique (Ritson, 2016; Yeo *et al.*, 2017). Finally, the openness and informality of the genre, which often entail the merging of public and professional discourse, promotes open participation in review on social media. It has been suggested that blogs and tweets facilitate extended review, that is, review not only by experts in the discipline (Trench, 2012: 278), and make it possible to harness 'collective intelligence' and the 'wisdom of the masses' (Minol *et al.*, 2007: 1132) in the review process: experts, researchers in other disciplines and the wider public can all participate to legitimize or delegitimize scientific information. This openness, however, also raises the question of the quality of open review, with some researchers rejecting the reliability of the 'wisdom of the masses' in the review process. Fahnestock (2016), for instance, argues that the internet can amplify uninformed responses to research reports. Controversy has emerged in the blogosphere and on Twitter about who has the right to critique published papers (the wide public or only members of the community) (Ritson, 2016) and the appropriate discourse for doing it (Buehl, 2016). For example, as Brunner (2016) reports, when a paper published by a group of NASA scientists (Wolfe-Simon *et al.*, 2011) was heavily contested and criticized on the blogosphere, NASA tried to discredit this criticism by calling the critics 'bloggers in pajamas'. However, the fact that most of the critics did not blog or tweet anonymously showed that most of them were active researchers.

Several studies provide evidence about the role of blogs and Twitter in the landscape of academic review genres (e.g. Buehl, 2016; Luzón, 2013; Mandavilli, 2011). First, these platforms contribute to democratizing the process of evaluating and legitimating knowledge. Unlike traditional forms of one-way post-publication peer review (e.g. printed letters to the editor, commentary articles), the openness and commenting capabilities of blogs allow expert debate to happen 'outside the reviewed publication system' (Bazerman, 2016: 275). They also foster the visibility of this debate, bringing science close to the public and promoting a culture of open science. New media can help researchers to overcome some barriers and limitations of formal peer review (e.g. disciplinary barriers, hierarchical barriers, temporal limitations) in different ways: by allowing cross-disciplinary discussion and interaction among researchers in related disciplines; enabling novice and less experienced or peripheral researchers to become part of the peer review system, and contribute to knowledge production and validation; and facilitating quick feedback, criticism and response on papers.

Blogs and tweets are also useful tools to bring flaws in published research to the attention of the academic community. Some blogposts, which we have called elsewhere 'research commenting posts' (Luzón, 2013), comment on published research by offering not only explanations but also evaluation and, in some cases, criticism. In our study of 'research commenting posts' from Researchblogging.com (Luzón, 2013), we found that, while many presented a positive stance toward the findings and thus contributed to promoting the research, almost 40% of the posts questioned some aspects of the research or refuted claims made in the paper. Similarly, in their study of posts that cited peer-reviewed articles in Researchblogging.org's Health, Shema *et al.* (2015) found that almost 30% of the posts included some criticism of the paper. The term 'take-down' is used by Mehlenbacher (2019b: 133) to refer to the type of post 'which reports flaws in a study, discusses a retraction, and may even report alternative analysis and findings'. She suggests that it may be a new genre, with antecedents in genres for communication between experts as well as in journalistic discourse.

Blogs and tweets may therefore play an important role in detecting questionable published research faster than traditional post-publication review genres and in forcing action from editors, as revealed by various controversies that began as a result of bloggers' critiques of papers published in peer-reviewed journals. The most prominent example is the controversy surrounding the article published online in the journal *Science* by Wolfe-Simon *et al.* (2011), who claimed to have discovered bacteria that could use arsenic instead of phosphorous to sustain their growth (Buehl, 2016; Mahrt & Puschmann, 2014; Yeo *et al.*, 2017). Rosemary Redfield, a microbiologist at the University of British Columbia, wrote a blogpost in which she reported on failing to replicate the findings of the paper herself and criticized the method. This post, while written in a relatively informal style, shares features with the reviews in scientific journals, such as technical vocabulary, figures, illustrations and references to support her points. After Redfield expressed concerns about the paper, discussion and criticism of the study developed through comments in her blog and those of other colleagues. These critiques were shared and discussed on Twitter, which contributed to increasing their salience and visibility to wider audiences (Yeo *et al.*, 2017). Formal traditional post-publication review took much longer to be available, in the form of eight technical reports and a 'News & Analysis' piece, published alongside the printed version of the paper, six months after the controversy. Another example of the public refutation of a published paper through well-argued criticism in a blog and the discussion in comments is discussed by Sidler (2016). Upon the publication of a paper with seemingly questionable claims in the *Journal of the American Chemical Society*, chemist Paul Docherty replicated the experiment and liveblogged it, posting figures and data in real time. The ensuing criticism and discussion in the comments led to the withdrawal of the paper. The implication Sidler (2016: 100) draws is that 'even publications in the most prestigious journals are vulnerable to the near-instant access and scrutiny afforded by the Internet', which shows again that the new digital genres are transforming knowledge production practices, and that knowledge is more than ever in a state of flux and always susceptible to being modified.

Informal post-publication review on Twitter can also complement traditional peer review by helping to identify problems with methodology or incorrect findings. Twitter was used by life science researchers to criticize the methodology of a paper whose authors claimed to have identified genes that could predict the human lifespan (Mandavilli, 2011). This criticism led to *Science* issuing an 'Expression of Concern' and to the retraction of the paper. A more recent case of a retraction because of criticisms of the methodology via Twitter is that of a *Nature Communications* paper that suggested that having female mentors had negative effects for the career of female scientists (Oransky & Marcus, 2020b). In an editorial accompanying the retraction, the editors stated that the publication of the paper and the criticisms that it attracted led them to reflect on their editorial processes and review their editorial practices and policies, and strengthened their 'determination in supporting diversity, equity and inclusion in research' (Regarding Mentorship, 2020: 1), which suggests that post-publication review on social media could have a role in changing reviewing practices. The fast feedback provided on Twitter is one of the main advantages of this type of review, as it can help to identify incorrect findings rapidly, and prevent them from being validated as knowledge which can influence other people's work (Mandavilli, 2011). Despite these benefits, post-publication comments on Twitter risk being misused by trolls and social bots in order to manipulate public opinion on scientific issues (Sharevski *et al.*, 2020).

With growing concern about academic integrity and an apparent increase in the number of retractions in recent years, 'watchdog blogs' (Gross & Harmon, 2016) such as 'For better science' (https://forbetterscience.com) are intended either to detect or expose error and fraud. The blog 'Retraction Watch' (https://retractionwatch.com/), for instance, brings retracted articles to the public's attention. The posts on this blog include information on the retracted paper and the reasons why it has been retracted, as well as testimonies of the parties involved, such as quotations from comments by the retracted authors and the editors, responses of the authors or editors to the bloggers' queries, in the rare cases when they get such responses, or links with additional evidence to support the bloggers' points. The language is informal and engaging, making use of strategies of journalistic discourse. Humor and irony are frequently deployed to shame the authors when the retraction is due to misconduct. As Bazerman (2016: 275) concludes, such blogs have a place in scholarly communication because they make it 'far more likely that serious accusations of fraud, plagiarism, other violations, and retractions do not get lost in the interstices of a loosely articulated communication system'. However, these blogs (as well as platforms like *PubPeer*) which alert readers to errors or misconduct in science in order to force change may also have a dark side, as pointed out by Teixeira da Silva (2016): a fair defense might not always be possible for all researchers, and public shaming, even for unintended errors, can destroy a researcher's reputation.

10.4 Research Case Study: Engaging in Scientific Controversies in Blog Comments

As we discussed above, the commenting capabilities of blogs and their open nature mean that academic blogs provide a new space where experts can evaluate others' arguments and claims and discuss controversial scientific issues. While the discussion may be triggered by a publication, commenters may go beyond the evaluation of a specific published paper to take positions on a significant scientific dispute or a public controversy. Comments offer blog readers the possibility of participating in long communicative exchanges where anybody interested in the topic can express allegiance or opposition to comments upholding a position. In polylogues where several participants interact in the comments section, a commenter may respond to one or several previous comments.

Thus, blogs may introduce new knowledge construction practices by enabling diverse and heterogeneous audiences to 'participate in co-constructing research debates' (Mauranen, 2013: 30–31). Researchers have documented how these debates are achieved by analyzing aspects such as dialogicity, intertextuality, interaction and evaluation in academic and science blogs (Bondi, 2018; Luzón, 2014; Mahrt & Puschmann, 2014). In this section, we report the results of our study of three controversies developing in posts through their comments (Luzón, 2014) in order to show the role of blogs as sites to discuss controversial issues and to illustrate the discourse features of these interactions.

The first controversy, developed in posts in the blogs *The Intersection* and *Aetiology*, was triggered by the publication of the book *The Oxford Book of Modern Science Writing*, edited by Richard Dawkins (2008). Of the 83 texts included in the book, Dawkins selected only three texts by women, leading female science bloggers to criticize the underrepresentation of female scientists. The second controversy was over the blog entry 'Should Boys Be Given the HPV Vaccine? The Science Is Weaker than the Marketing' (in the blog *The Crux*, hosted by *Discover* magazine), where Jeanne Lenzer, a medical journalist, claimed that there was not solid clinical evidence to support the administration of the HPV vaccine to boys. In response, Tara Smith's post on the blog *Aetiology* refuted Lenzer's claim, triggering a discussion about the benefits and cost-effectiveness of giving the HPV vaccine to boys. The third controversy related to climate change, and was triggered by a study on the impact that the changes in clouds may have on climate. Supporters of different positions defended their views and attacked others' claims on the blogs *Real Climate* and *Roy Spencer's blog*. These posts and related comments were part of broader disputes between supporters and critics of opposing views on these three issues.

The corpus of posts and comments (6 posts and 50 comments per post) was analyzed for how participants in the discussions evaluated other's claims and ideas or expressed allegiance to a group or their support of a particular position. For that purpose, we identified and coded indicators both of social or positive relational behavior and of rude or antisocial behavior, using a coding scheme developed by Luzón (2011). This coding scheme was based on research on social presence in computer-mediated communication (Rourke *et al.*, 1999), interpersonal interaction in academic discourse (e.g. Hyland, 2000) and uninhibited behavior and impoliteness (e.g. Culpeper, 1996), as well as on indicators emerging from the data. We identified three categories of indicators of social presence: *affective indicators* (i.e. personal expressions of emotion and feelings), *cohesive indicators* (i.e. verbal behaviors that convey a sense of group commitment and community) and *interactive indicators* (i.e. indicators providing evidence that the other is willing to contribute). Similarly, we identified three categories of indicators of rude or antisocial behavior: *indicators of negative socioemotional behavior* (i.e. personal expressions of negative emotions or feelings), *indicators of group exclusion* (e.g. refusal to consider the other as a valid member of the group) and *indicators of confrontational interaction* (i.e. interaction intended to seek confrontation). Tables 10.1 and 10.2 list the different types of indicators of social and antisocial behavior, respectively.

As can be seen in Tables 10.1 and 10.2, indicators of antisocial behavior were more frequent than indicators of social behavior (1.84 vs. 1.26 occurrences of indicators per post). The most frequent indicators of cohesive behavior were inclusive pronouns (indicating cohesion) (137 occurrences) and collaboration (indicating willingness to contribute) (77 occurrences). Inclusive pronouns in the corpus appear to be used to remind readers of shared knowledge and thus bring them to the writer's interpretation and opinion, or to refer to a narrower group, such as other

Table 10.1 Indicators of social behavior

Indicators of social behavior (378 occurrences)	
Affective (44 occurrences)	Paralanguage, such as emoticons or parenthetical metalinguistic cues (e.g. 'Ohhhh!')
	Expressions of oral discourse (e.g. 'I dunno')
	Verbal expression of emotion
	Affective use of humor
	Self-disclosure
Cohesive (186 occurrences)	Vocatives
	Inclusive pronouns to refer to the group
	Appeal to shared knowledge
Interactive (148 occurrences)	Acknowledgement (i.e. quotation or referring to another participant's message)
	Approval (i.e. complimenting, thanking)
	Agreement or polite/hedged disagreement
	Collaboratively following the thread (e.g. answering questions, giving further information)
	Requests/invitation to respond

Table 10.2 Indicators of antisocial behavior

Indicators of antisocial behavior (533 occurrences)	
Negative socioemotional behavior (59 occurrences)	Non-verbal expressions of negative emotion (e.g. emoticons, capitalization)
	Irony, sarcasm
	Verbal expression of emotion (e.g. 'I'm tired of', 'very disappointed')
Group exclusion (223 occurrences)	Vocatives, to identify the target or attack or criticism
	Excluding 'you', to contrast with 'we' (in-group)
	Personal attacks, insults (e.g. 'Unlike you, I'm sane')
	Disassociate with the other
Confrontational interaction (251 occurrences)	Argument criticism
	Quoting and referring, for mocking or strong criticism
	Disagreeing/correcting
	Imperatives/directives used for confrontation

members of the writer's discipline (see Example 10.1a) or to participants sharing the writer's opinion (see Example 10.1b).

Example 10.1

(a) With every other vaccine-preventable disease, *we* look at reduction in both morbidity and mortality – why is HPV being treated so much differently than, say, chickenpox, pneumo or Hib? (*Aetiology*)

(b) They are *our* kids, not *yours*. *We* will decide what is injected into their bodies, not *you*. (*The Crux*)

Collaboration consists in supporting one position by contributing further to the issue introduced in the post, that is, adding further information, providing references and links that could be useful for the discussion and answering questions (see Example 10.2).

Example 10.2 Nice summary. I do some research in the area, specifically oral HPV and head and neck cancer, and have a few points to add. (*Aetiology*)

Indicators of social behavior are sometimes combined in the post with indicators of antisocial behavior toward participants holding the opposite position. This is clear in the opposition between 'our/we' and 'yours/you' in Example 10.1b or in Example 10.3. The expressions of agreement ('I'll second dyson's comments. ... I agree with dyson') are combined with argument criticism to express opposition to the views of the other group.

Example 10.3 I'll second dyson's comments. ... I agree with dyson that many of your arguments are cherry picked and ideological. (*The Crux*)

As pointed out above, indicators of antisocial behavior are more frequent than indicators of social behavior, including not only disagreement, correction or argument criticism, but also acts that are less common in academic culture and discourse, such as irony/sarcasm, personal attacks or insults. Irony is used by participants to refute others' arguments through ridicule, thus construing consensus among those that are 'in' to the ironic comment (Attardo, 2001). For instance, the comment in the thread about the low number of women in *The Oxford Book of Modern Science Writing* shown in Example 10.4, which responds to the blogger's criticism that Dawkins should have put together a more balanced anthology by including more essays written by women, sarcastically suggests that expecting a quota for any group is nonsensical. The feigned intimacy and surprise, conveyed by the direct address to the reader ('did you notice...') and the interjection 'gee', emphasize the irony and the condescending tone.

Example 10.4 And where is the balance of black vs. white scientists, of Buddhists and Protestants, of 20-somethings and 60-somethings, of Russians and Americans; and did you notice neuroscientists were almost entirely shut out (gee, and no botanists, I guess plants have no importance)? (*Aetiology*)

Personal attacks and insults occur surprisingly frequently (81 occurrences) for a discussion that includes working scholars. Participants seem to seek to undermine the credibility of those holding opposite views by questioning their intelligence, knowledge of the topic or objectivity (see Example 10.5).

Example 10.5

(a) @32- This is where reading comprehension would pay off, instead of just making things up to get defensive. (*Aetiology*)
(b) My question, how can a working scientist like *name* be so ostensibly thick that he actually reads 'clouds cause ENSO' into those emails? I don't get it. {:-| (*Spencer's blog*)
(c) Excuse me, have I just walked into a Kindergarten class? (*Spencer's blog*)

The category of indicators of antisocial behavior with the highest number of occurrences (90.3 per 100 messages) is *Indicators of confrontational interaction*. Participants attacked others' position by negatively evaluating their claims, arguments, ideas or theories (e.g. references to false arguments, flawed analysis) or criticized opponents for using logical fallacies in their arguments (see Example 10.6) or for ignoring important facts and thus deriving wrong conclusions.

Example 10.6 First, your attempt to compare the abortive attempts to do carotid artery bypasses to the HPV vaccine trials is comparing chalk and cheese. (*The Crux*)

Participants also harnessed the 'cut-and-paste' feature of the medium to quote fragments of previous comments with which they disagree or that they want to refute, in order to contextualize their response and present an antagonistic exchange in a single message. These discussions are polylogues, so quoting or referring to previous comments helps participants make clear whose argument they are refuting or criticizing.

The analysis of the indicators of social and antisocial behavior reveals that the interlocutors frequently use comments to signal their allegiance to a particular group by construing conflict with those who support opposing theories/ideas. In order to refute the validity of others' theories and ideas, participants resort not only to scientific arguments and strong evidence, but also to other strategies of weak argumentation (e.g. sarcasm, personal attack of commentators with opposing ideas).

The results of this study offer insights about the nature of blog discussions of controversial scientific issues and their relation to other forms of communication among researchers. These discussions are related to other highly critical and confrontational scholarly genres such as the medical journal editorial or the conflict article, a genre in which 'the writer explicitly opposes a named other writer or paper' (Hunston, 2005: 1). As in these genres, interlocutors in blog disputes defend their claims by engaging in adversarial arguments and resorting to unmitigated criticism and more explicit and personal criticism than is common in other academic genres (e.g. the research paper). However, controversial discussions in science blogs have distinctive features, which derive mainly from the affordances of the medium. Confrontational exchanges through traditional printed genres tend to be a two- or threefold exchange (i.e. the criticized article, the response/conflict article and the response by the author of the criticized article), with few interlocutors taking part. By contrast, in blog discussions conflict is constructed cumulatively along several exchanges, interactively and collaboratively, with several interlocutors taking sides and responding to previous comments. Furthermore, blog discussions are polylogues, where each comment can respond to the post or to one or various previous comments and where interlocutors can answer back as soon as the others' comments are published. In addition, even if most participants in the blog discussions analyzed seem to be researchers or people aware of academic culture, they sometimes move beyond conventional academic politeness and adopt discursive strategies that are less frequent or non-occurring (e.g. challenging questions and coercion, insults, curses) in printed antagonistic genres, such as the conflict article or book review. Again, these features suggest that the open participatory framework which facilitates the participation of interested readers with various degrees of expertise and the anonymity allowed in the comments may also pose problems, such as destroying the reputation of authors or bloggers, making irrelevant, uninformed or tangential comments which may hide well-supported comments, or, as Fahnestock (2016) points out, the prolongation of controversies. In addition, as Bloomfield and Tillery (2019) have cautioned, digital genres make it easier to circulate non-scientific information and to reinforce and validate communally skeptical perspectives about, for example, environmental issues.

10.5 Concluding Remarks

This chapter has shown that the internet is transforming reviewing and quality evaluation practices with new affordances and limitations. The affordances facilitate the emergence of new genres (or adaptation of old genres) for the evaluation and legitimization of research which are compatible with and complement traditional peer review genres. These new genres respond to the demand for more open and participatory review processes: they make it possible for various reviewers, not only those selected by the editors, to participate in the evaluation of research; they make the process more transparent, by facilitating the disclosure of the reviewers' identity; and they contribute to presenting the research reported in published papers as 'open' to evaluation, which might eventually lead to changes in the research article genre.

In terms of pre-publication review, one useful innovation is the collaborative review forum, where the interaction between editors, reviewers and authors can accelerate evaluation and publication. Other practices, such as disclosing the authors' or reviewers' identities, pose greater challenges, i.e. authors and reviewers can be reluctant to submit or review papers, as it could affect their reputation; reviews could be less honest, particularly if the reviewed author is in a position of power. A practice adopted by some journals to solve these problems has been to make optional the disclosure of identity.

In the post-publication review process, the changes that these genres may bring about are particularly noteworthy: post-publication commentary may be used by authors to improve their papers, but authors may also participate in the discussion to clarify their claims and respond to criticisms. In addition, controversial and disputable claims in published papers are subject to public scrutiny and open to discussion by a large pool of reviewers. The open commentary and public discussion of published research enabled by digital genres seems to be a positive innovation which can give voice to underrepresented scholars and to the interested public, thus contributing to the democratization of science and to the promotion of a culture of open science. However, as we have discussed in the chapter, it also has its perils (i.e. trolling). Of particular concern is that, as Fahnestock (2016) notices, the internet may amplify uninformed responses to research from commenters without the requisite expertise, which may attenuate the value of some published research, and contribute to the dissemination of disputed claims.

11 Final Considerations and Future Directions

In this book, we have aimed to show how new social demands and technological advances have driven the emergence of digital genres for knowledge production and communication and how these new genres both reflect and facilitate broader changes in academic communication (i.e. new purposes, new audiences, new production practices). We have discussed the multifarious possibilities for rhetorical action provided to researchers by the expanding repertoire of digital genres and have described some of these genres in more detail. Through the use of research case studies, we have illustrated some of the digitally mediated practices scholars can engage in today. These practices are both afforded and constrained by the medium and are the result of new rhetorical needs, such as increasing researchers' visibility and impact or opening research results to access by both professionals and the interested public. While the description of digital genres in this book is not comprehensive, we hope it provides evidence of the new dynamics that facilitate the production and dissemination of academic knowledge using Web 2.0. In this closing chapter, we summarize the main findings of our analyses, discuss implications for genre theory, genre analysis and genre pedagogy, and sketch out possible areas for future research on digital genres and multilingual scholars' communication practices.

11.1 How Academic Communication is Changing: New Affordances, New Discourse Practices and New Challenges

Drawing on the conceptualization of genre as a semiotic tool for accomplishing actions, this book has provided an overview of the range of opportunities for the production and dissemination of knowledge that scholars find on the internet. The chapters have illustrated that researchers can 'show' their research through audiovisual genres, share their research in progress with peers, and assess and discuss it in new ways. They can also interact with peers on academic social networking sites (ASNSs), disseminate their work to diversified audiences and engage the public in research processes. We have analyzed the way that scholars construct multiple identities through their discourse choices in various genres and platforms, and how they make meaning by combining the various semiotic modes made available in specific genres. We have also explored the languages that multilingual scholars deploy in these new discourse practices and have shown that being able to choose and combine languages on the internet is an affordance that enables them to reach multiple audiences. Broadly speaking, the discussion and analyses of digital genres in this book have confirmed: (i) the rapid proliferation of genres, facilitated by technology, to meet specific academic needs – what Swales (2004: 4) refers to as 'generification' of academic life – with myriad digital genres emerging, evolving and diversifying into subgenres (e.g. the different types of blogs produced by scholars); (ii) the key role that these digital genres currently play in working with other genres to support research activity and academic knowledge communication both within and beyond traditional discourse communities; (iii) the range of possibilities for communication with diverse audiences, collaboration, visibility and evaluation offered by online genres as a result of the affordances of the web (e.g. multimodality, hyperlinking, reusability, modularity, wide reach, interactivity); and (iv) the concomitant changes in scholars' communicative practices, as a result of the need to adapt their communication strategies to meet new academic expectations, to achieve new purposes and to reach diverse audiences.

Beyond the affordances of Web 2.0 technologies, the emergence of digital genres for knowledge communication has also been driven by new socioeconomic demands and related changes in research culture, that is, researchers in some disciplines are increasingly expected to respond to new exigences related to open science culture, and may need or wish to make their research visible and accessible at international, national and local levels. In many disciplines, research is a group or collaborative activity, and is becoming more interdisciplinary and globalized, thus involving researchers from different linguacultural backgrounds. Throughout the book's chapters, we have discussed new and diverse digitally mediated knowledge communication practices, which help researchers meet demands for visibility, collaboration, openness and reproducibility. Firstly, we have seen that digital genres help to share research and to maintain collaboration, above all, international collaboration, and to seek connections and informal networking. A number of digital genres have emerged to support communication with peers, for example, to share research findings in various formats and to disseminate detailed information on research methods and procedures (e.g. open lab notebooks [OLNs] and video methods articles [VMAs]). Providing this detailed documentation of research practice contributes to the reproducibility and reuse of research output and thus to research efficiency.

Secondly, we have shown that digital genres offer a wide range of possibilities for scholars to make their research work accessible not only to expert audiences (peers) but also to other audiences with various degrees of expertise, contributing both to more participatory knowledge construction and to public understanding of and engagement in science. Complementing formal genres for expert-to-expert communication, new genres have emerged that enable scholars to establish an informal dialogue with their peers and the interested public. The technological affordances assist such dialogue and the exchange of views and, in the case of public engagement in science (e.g. citizen science projects), further support participation in the production of academic knowledge beyond specialized (disciplinary) communities.

Thirdly, we have also explored how the affordances of Web 2.0 for openness, interaction and wide reach have enabled new practices of research evaluation. Traditional genres of peer review have evolved, facilitating a more participatory and transparent evaluation process, and new genres for the evaluation and discussion of published research have emerged (e.g. notes, comments), which may contribute to an ongoing conversation that improves the quality of research output. The awareness that the traditional evaluation system, in disciplines where it is based on publication in high-impact journals, can have negative consequences for researchers and for the advancement of research has led several hundred institutions and organizations, including funding bodies and publishers, and thousands of individual researchers, to sign *The Declaration on Research Assessment* (DORA). DORA emphasizes the 'need to improve the ways in which the output of scientific research is evaluated' (https://sfdora.org/read/), which involves evaluating the various types of research output (not only publications), eliminating the use of journal-based metrics of scientific quality and finding other ways to assess the quality and real 'impact' of research outputs. We have shown in this book that digital genres are useful tools to facilitate this type of evaluation.

The findings presented in this book also provide evidence for the claim that digital genres play a key role in the construction of academic identities. These identities are performed and negotiated not only in digital genres with a clear promotional purpose (e.g. various types of blogs and academic homepages) but also in genres for establishing and maintaining formal scholarly exchange (e.g. open access reviewers' reports), informal exchange (e.g. tweets) and collaborative work (e.g. OLNs) with other expert peers. Digital genres make it possible for scholars to construct multiple identities in order to meet the expectations of various audiences. In all of these genres, scholars position themselves not only within their communities, but also in relation to diverse public audiences. The explorations of genres at a discourse level also suggest that issues of authorial identity and visibility are key to digital genres. In the new genres used for discussion and evaluation of research, interpersonal features of discourse reveal how scholars make themselves visible, express criticism, openly convey their views (which may contradict others' views) and display various clines from adherence to detachment from other scholars' claims and arguments.

Lastly, the analyses reported in this book show how digital genres can harness the multimodal affordances of the medium. We have illustrated the way digital genres, ranging from blogs and microblogs to crowdfunding science projects and popular science videos, draw on multimodality as a key composing resource to attain their creators' rhetorical goals. Each digital genre uses a distinct configuration of semiotic resources in order to perform specific rhetorical actions. In genres mainly aimed at communicating science to lay audiences and engaging citizens in scientific research, various semiotic resources are combined both to increase the general public's science literacy and to persuade it to take action. In genres oriented to peer communication (e.g. graphical abstracts, VMAs, enhanced online articles), the visual mode offers the possibility to provide further details and a more accurate understanding of the contents of the genre and the context in which research was conducted.

The descriptions of digital genres in several of the book's chapters also reveal changes in discourse practices as compared with traditional genres: the rhetorical and lexico-grammatical features of these genres differ in some ways from those of formal genres of academic communication. This shift results from the need to achieve new purposes and accommodate wider audiences, and from the possibilities for multimodal and multilingual communication offered by digital media. For example, engaging the public in research processes, appealing to the public for responses in citizen science and crowdfunding genres, discussing research in progress, evaluating research openly or promoting one's research in Twitter require scholars to use different strategies from those used in traditional research genres. In some digital genres, scholars recontextualize specialist knowledge in order to communicate it effectively to audiences with different knowledge backgrounds and to engage these heterogeneous audiences. The discourse needs to be accessible and, above all, highly interpersonal and dialogic so as to prompt a response on the part of the intended audiences. This is perhaps the reason why the discourse of some of these genres exhibits a recurring use of the first- and second-person pronouns 'I'/'you', direct appeals to the audience in the form of questions, directives, attitudinal language and explicitly evaluative language. These are features of a more informal and involved style, found not only in genres to communicate with a wide audience but also in some digital genres for communication within the academic community, where the use of a personal style creates proximity with the readers (e.g. OLNs, ResearchGate Q&As).

These results contribute to answering the question of whether academic writing is becoming more informal. In their analysis of a corpus of research articles written over the last 50 years, Hyland and Jiang (2017) found a small increase in informality, although not in all the disciplines they studied. The results of our research tell a different story, one which reveals that, even if in research papers the shift toward informality may be slow, informal features are salient in many of the digital genres that scholars compose. Hyland and Jiang (2017: 47) associate the gradual increase in informality in research papers to 'changing rhetorical practices' intended to involve readers and claim solidarity with them and to a wider shift 'toward more engaged and intimate forms of interaction'. The digital genres analyzed in this book reveal the increasing importance of engagement in knowledge communication, and show that engagement is realized not only through linguistic forms such as rhetorical questions or appeals to the audience, but also through the various semiotic resources made available in the genre, such as emoji, pictures of researchers or familiar sounds and images in videos. This finding is related to the fact that academic communication online has become highly multimodal, with visuals being a key element in many digital genres, where they may be used, among other purposes, to provide evidence, to facilitate understanding or to attract and engage the audience.

Scholars' online communication has also become more multilingual and apparently more tolerant of non-standard linguistic forms when English is used as the shared medium of communication. We have discussed the issue of multilingualism across the book's chapters, providing reasons to suppose that digital genres may both facilitate participation of multilingual scholars in global academia and help us reach local communities and wider publics. As our exploratory analysis of genres has shown, many scholars' interconnectedness for the sake of advancing knowledge draws on English, which is being used as a shared language more than as a native language, which, echoing Seidlhofer (2011), can act as a tool that enables the participation of scholars from different linguistic backgrounds. Digital genres may offer more opportunities for multilingual scholars to disseminate their research internationally and to discuss and evaluate others' research and ideas, as in most of these genres there are no gatekeepers judging the use of English. Whereas in the domain of formal publication the question of whether 'English nativeness' makes a difference in getting research published still remains a topic of scholarly debate (Politzer-Ahles *et al.*, 2016), in some of the new genres described in this book 'native' usage is not always such an important concern. For many multilingual scholars, using English as an additional language in informal communicative exchanges on the web entails fewer burdens and less pressure than writing in English for research publication purposes. More attention is paid to maintaining communication rather than language correctness. In addition to communicating globally in English as a lingua franca, multilingual researchers also communicate and disseminate their work online by using the languages spoken in their local or national settings. Digital genres also offer the possibility of engaging in multilingual practices, and thus reach diverse audiences, by making it possible to alternate various languages (e.g. in research group blogs or Twitter) and to hyperlink to online texts in other languages. As explained in Chapter 7, one of the motivations to write posts or tweets in local languages is to make it easier for the local community to understand scientific research that has a strong local focus.

However, despite the new affordances of the digital environment for multilingual communication, digital genres do not seem to be dislodging English as the *de facto* lingua franca. In other words, in the online environment, users of English continue to relegate other languages to local (and one might argue lesser, subordinate) functions. Thus, we wonder whether some technological affordances are being purposely overlooked or perhaps underexploited. For example, machine translation (see Bowker, 2002) is a technological affordance that has been underexploited when using digital genres; the possibility of having parallel texts in different languages is not usually considered either by scholars or web developers. In addition, digital genres used by multilingual scholars to communicate with the local audience in their local language often have less value for evaluation purposes and for measuring the impact of research. In this sense, despite the multilingual practices of scholars on the internet, Lillis *et al.* (2010) claim that knowledge measuring revolves around English may still hold true (see also, more recently, Nygaard & Bellanova, 2018).

In our discussion of digital genres, we have also emphasized that although digital media and their affordances offer new opportunities for scholars, they also present challenges for them (particularly for peripheral scholars). Firstly, the openness, wide dissemination and enhanced visibility of research can help scholars build an academic reputation, but also involve tensions and potential perils, such as the dilemma of whether to devote research or personal time to social media activities, the lack of control of what information is visible to particular audiences and concerns about how unintended audiences can use such information. Secondly, the open discussion and evaluation of a scholar's work afforded by digital genres, together with the anonymity that some of these genres enable, also means that scholars are more exposed to trolling or to being publicly criticized, even by non-experts. In addition, the open participation in knowledge construction and discussion that digital genres facilitate comes into conflict with certain deeply ingrained values of research culture, such as the role of authority in knowledge validation.

Digital genres also impose some media constraints on knowledge production and circulation, which may be particularly daunting for marginalized scholars who may lack the technological infrastructure, material resources or the digital literacies to produce or access online texts. We would like to raise concern for the situation of peripheral scholars with limited resources and limited or no access to the internet, who may find it difficult if not impossible to communicate online with other members of their disciplinary communities. Canagarajah (1996) and Salager-Meyer (2008) remind us that, in less digitalized contexts, material resources (e.g. funding for research and internet access) often entail a greater burden on periphery scholars' participation in knowledge production than do linguistic resources. If scientific knowledge is increasingly produced, accessed and shared online, scholars who cannot take advantage of online resources and tools may be excluded from equitable participation in global academia. Thus, when it comes to the opportunities for knowledge production and dissemination that digital genres offer, the question of power is highly relevant: who has the infrastructure and resources not only to get access to knowledge, but also to produce digital genres, to make their research visible and thus gain recognition, to lead international collaborations or to decide what issues are worth discussing online? As scholars in the Global North are increasingly taking advantage of digital genres and online platforms to promote their research and make it searchable, visible and accessible, digitalization may result in more inequality in global academia.

We also envisage conflicting interests resulting from the affordances of digital technologies for easy and quick access to research results and publications, for example, between open science or open access and the various economic interests at play (e.g. paywall journals, for-profit ASNSs). We have discussed the tension between the open sharing of scholars' work encouraged and enabled by ASNSs (such as Academia.edu) and the for-profit nature of some of these sites. Although these ASNSs provide the infrastructure for a model of knowledge building based on free access to publication, open sharing and participation, they also make a profit, thus contradicting the principles of open science. New genres make it possible to share research (e.g. OLNs, pre-prints), overcoming the problem of delays in journal publication. The increasing number of genres and platforms to disseminate research may lead to changes in the traditional publishing model, where academic service providers control access to content.

11.2 Implications for Genre Theory and Genre Analysis

Throughout this book, we have aligned with Miller's (1984: 163) seminal view of genre as a conventional category of discourse that fulfills rhetorical and social actions. This view has proved apposite for understanding the complex communicative goals and rhetorical exigences met by digital genres and reflected at a discourse level. We have also emphasized that digital genres have features which result from the affordances and constraints of the medium and that should be taken into account when analyzing these genres.

We have illustrated that a distinct feature of most digital genres is their hybrid, non-linear, fluid and hyperlinked nature, which means that the boundaries between genres are not clear-cut in the digital environment. The production of digital genres is often the result of assemblage, reuse and hyperlinking, that is, of establishing interconnections. As shown in the book's chapters, there is considerable genre embedding (i.e. genres within genres, such as online research articles containing author videos) and linking (e.g. academic homepages or blogs linking to online research articles). The modularity affordance of the internet facilitates the production of what we have called 'nested genres' (see Chapter 5), such as the posts that make up an OLN. The various tweets in the Twitter account of a research group can also be considered nested genres: while each has a specific purpose, they all contribute to the overall purpose of promoting the group's research. Many of the genres we have discussed could be regarded as 'open' or 'living' genres, as new content can always be added or existing content can be changed and updated. This is the case of OLNs, used by researchers to discuss research in progress by adding new posts, Q&A threads in ResearchGate or posts in the research group blogs. The possibility for updating and adding content offered by new media is particularly relevant for genres for knowledge production and communication, because it makes it possible to present knowledge as always open to revision and research as an ongoing process (see, for instance, the example of living reviews). All these features should be analyzed when exploring digital genres for knowledge communication.

This book has also provided insights into genre evolution and innovation. In Chapter 2, we noted that Miller and Fahnestock (2013) call for more research to answer several questions regarding the emergence and evolution of genres. Based on insights presented throughout the book, we can assert that it is new social exigences that account for the emergence of new genres or the evolution of genres in the digital environment. Digital genres enable scholars to give visibility to their research, disseminate science to a wider public, enhance the validity of their methods and accomplish rhetorical exigences for the transparency and rigor of their research protocols. Examples of genres that have emerged to meet such new exigences are VMAs, which emerged from the need to help other researchers reproduce experiments with fidelity, and combine visual and verbal modes to respond to this need; citizen science projects, which engage the general public in scientific research processes (data collection, data classification, and analysis); or crowdfunding projects, which have evolved from other genres, such as project proposals or business plans, to respond to the need to make science accessible to society in general and obtain funding for research from the general public. An important quality of this emerging repertoire of genres for knowledge communication and dissemination is therefore their versatility or capacity to respond to the new social exigences driven by current socioeconomic and research policy agendas.

All the digital genres discussed in this book appear to have printed or digital antecedents. While they derive from existing genres and adopt some of their features and conventions, they incorporate functionalities that make them different from their antecedents. In fact, some of the genres we have discussed have evolved significantly, harnessing the affordances of the web to respond to new rhetorical needs (e.g. crowdfunding projects, OLNs) to the point that it is difficult to determine whether they are an evolution from their antecedent genres or, as Wickman (2016: 16) puts it, a 'complete break' from them. What seems clear is that, as genres are 'social actions' (Miller, 1984), it is the action(s) performed by digital genres that will ultimately help to determine whether they are different genres from their printed counterparts.

Although dynamism is an inherent feature of all genres (Berkenkotter & Huckin, 1995), evolution and innovation, driven by new social needs and demands, are particularly important to understand digital genres. Generic hybrids, such as crowdfunding projects, video abstracts or VMAs attest to generic innovation, evolution and ongoing change enabled by the technological developments of the internet. For example, Spicer (2014: online) describes video abstracts as 'a natural evolution of science communication into multimodal environments'. They direct expert readers' attention toward scientific content that is further elaborated in the online paper, but some of them are intended to attract a broader readership. VMAs have evolved from the traditional methods section by using audiovisual narration to enhance the contents of their associated online verbal texts. If the digital genres discussed in this book prove functionally effective for the members of the communities using them, they are likely to stabilize and eventually become part of the repertoire of genres deployed by scholars in the future. Considering the unprecedented generic innovation and evolution of online academic genres to date, it is not difficult to expect further innovation and new digitalization changes in scholarly communication practices. And while we witness the emergence of new genres and their stabilization in some cases, for example blogs, it is also feasible that some other genres will lose momentum, as was the case of AudioSlides (see Pérez-Llantada, 2021b), or may disappear because the community does not recognize them as useful or supportive of their professional communication needs.

This discussion has important implications for genre analysis. First, the analysis of how producers of digital genres achieve their purposes and meet social exigences cannot be restricted to the verbal mode or linear text. The discussion and analysis of genres in this book supports van Leeuwen's (2005: 74) claim that genre analysis has to be expanded 'when we transfer it from the domain of "monomodal", linear texts to multimodal, nonlinear texts'. This involves not only analyzing how various semiotic modes are orchestrated to achieve rhetorical purposes, but also accounting for the complex and varied relations between genres that can be established in the digital environment. In addition, since digital genres are the result of genre evolution and innovation, analysis could also involve comparison with their (pre-digital or digital) antecedents, in order to determine how they transform existing genres for academic communication and what new practices researchers engage with when producing them.

11.3 Pedagogical Implications: Widening Scholars' Genre-Based Practices

In addition to the implications for genre theory and genre analysis, important implications can also be derived for the field of English for Academic Purposes (EAP) and genre pedagogy. The increasing use of digital genres to communicate research both to peers and the public, and their possibilities for complying with the principles of open science might suggest that it is necessary to redefine successful knowledge communication in the 21st century and provide instruction to help scholars (including multilingual scholars) develop academic literacies and become effective communicators. From our discussion and analyses of digital genres, it is possible to set out some general recommendations for pedagogical practice. For example, given a likely increase in genre diversification as a result of new social demands, it is important that EAP instruction draws attention to scholars' emerging communication needs and raise awareness of the technological possibilities offered by digital genres and of the specific affordances and constraints of each genre (Pérez-Llantada, 2021b). Next, as we have argued elsewhere, formal instruction should offer rich input and opportunities for production of the recurring discourse features that characterize these genres, to make scholars aware of how these conventions contribute to achieving specific purposes (Pérez-Llantada & Swales, 2017). Instruction should pay attention to features of language and discourse style (e.g. degree of formality, linguistic and rhetorical features), interpersonal features (e.g. to create proximity with readers, to express authorial stance and evaluation), multimodal features and aspects of voice and identity construction. Another key practice that genre-based instruction can support is using strategies for the recontextualization of disciplinary knowledge, as it is through these strategies that scholars can engage effectively with non-expert audiences (otherwise knowledge remains circumscribed within the scholarly community). Teaching scholars to create add-on genres to enhance research articles (e.g. highlights, graphical abstracts) is also important, as is adapting academic knowledge to different genres, modes and audiences.

Special attention should also be paid to scholars' development of digital literacies (i.e. the skills needed to communicate, create meaning and relate in digital media) (Jones & Hafner, 2012), and multimodal composing skills (Hafner & Ho, 2020). In this book, we have shown that digital genres are tools to *perform identities*, *relate* to audiences within and beyond the community and *produce* and *communicate* knowledge in new ways. Thus, it is important to provide instruction that helps scholars use these genres to manage impressions; be visible and attract the attention of diverse audiences; establish and manage relationships; and combine semiotic resources strategically to achieve specific purposes. In agreement with previous research (e.g. O'Halloran *et al.*, 2016), we argue for developing multimodal composing skills so that scholars can learn about the possibilities for interaction and integration of modes in online genres. We would propose extensive practice in multimodal composing for a variety of purposes and audiences, as well as guided practice focused on digital skills in the effective creation of audio-visual (spoken and written) texts and the treatment of images, photographs, animations and other types of visuals for different rhetorical purposes. Digital literacies also involve awareness of the possibilities that the digital environment offers for reputation building and identity construction and of the need for careful management of this online presence. Scholars can be taught how to connect different online spaces to enhance the visibility and dissemination of their work.

Instruction should also inform learners about new possibilities for multilingual communication through the use of social networking sites and social media networks. To take advantage of the benefits of digital genres, multilingual scholars need to shift languages and registers to address different audiences. Tasks for practicing how to engage in formal and informal conversation with expert and public audiences can raise awareness of the benefits that users of English as a lingua franca can find through these media to share research and receive feedback. In addition, these tasks can be pedagogically useful to demonstrate the benefits enjoyed by multilingual scholars for targeting local communities through the use of the local language(s), and to make multilingual scholars aware of the value of our local language(s) to communicate in specific contexts. In previous chapters, we have highlighted the ways in which an expanding repertoire of genres has great potential for disseminating academic knowledge locally and globally and underscored that multilingualism is an advantage that enables scholars to reach both local and global audiences. Given these findings, in this chapter we underscore the value of multilingual web-mediated science communication and the need to support multilingual communication as an opportunity for knowledge dissemination.

11.4 Areas for Future Research

In this closing section, we suggest some areas for future research. From a genre analytical perspective, it would be helpful to see cross-pollination in future studies on digital genres for knowledge production and dissemination, for example, Swales's notion of communicative purpose and Miller's emphasis on exigence. Both have proved to be crucial conceptualizations to comprehensively describe the genres we have analyzed in this book. There is also a need for in-depth analyses of the genres we have explored in this book as well as comparative analyses with other digital or printed genres. These studies could provide clearer insights into the semiotic work of these genres (e.g. textual and visual features), their rhetorical properties and, more broadly, the degree of innovation, hybridization and change in and across interrelated genres in digital environments. The studies presented in the book's chapters point to the need for further enquiry into all these aspects. Further, the scope of this research can be extended to multilingualism and language choice in digital genres, for example by looking at genres in one language with embedded genres in a different language or languages or, as another example, by examining genres in one language that are hyperlinked to genres in another language or languages. Research on multilingual genres and multimodality, with a focus on visual rhetoric, could better inform the design of pedagogical approaches and the creation of multilingual and multimodal materials.

Researching digital genres also offers possibilities to address in further depth issues of language and multilingualism. For example, examining features of texts produced by users of English as an additional language could help define the scope of linguistic and cultural diversity that characterizes the use of English across digital contexts (Mauranen, 2018). In addition, more research is necessary on the extent to which multilingual scholars are producing and participating in digital genres in English and in local language(s), and on the experiences and attitudes toward participation in digital communication of scholars from different geopolitical locations. Research on the experiences of multilingual scholars when writing for publication has revealed that writing research in English can be more challenging for them than for English 'native' speakers (Hanauer *et al.*, 2019), which may result not only in the under-representation of multilingual scholars in international journals (Yen & Hung, 2018), but also in a reduced use of social media to communicate research in English. It is also unclear whether and how the absence of gatekeeping in many digital genres and changes in the dynamics of pre- and post-publication practices will affect the participation of peripheral scholars in global knowledge production and in the legitimization of scientific knowledge. More research is needed to critically address whether the processes enabled by digital media (e.g. sharing of data and publications, international collaboration) result in a more inclusive model of knowledge production. This issue goes beyond the scope of genre analysis but is worth investigating further.

Another aspect on which more research is needed is the reception and uptake of these new digital genres. Further research is necessary on whether digital genres that target interested audiences, including professionals and practitioners in the field who could benefit from the knowledge disseminated, are being taken up by these imagined audiences and how to compose para-scientific genres so that they effectively reach diverse audiences. This research would contribute to defining and assessing real research impact and to determining which genres are more appealing to particular audiences. Similarly, research on how related genres intended to enhance research articles (e.g. graphical abstracts) are perceived, accessed and used by scholars would help researchers, authors and journal editors decide whether these genres are worth producing.

So far, we have proposed lines of research for examining the affordances of digital genres for knowledge communication and dissemination. Yet, a critical look into their limitations is also necessary. We can identify at least three possible limitations that suggest corresponding research questions. The first concerns discipline. Are the existing affordances of the internet equally useful to all researchers across disciplines or mainly to those working in, for example, science, technology, engineering and mathematics (STEM) fields? In other words, might the nature (or *ethos*) of the discipline hamper scholars' engagement in certain genres thus making some disciplines more favored than others? The research literature reveals important gaps in our understanding of how digital genres are deployed for communication of knowledge in the humanities and the social sciences to the general public. A relevant research goal would be the examination of the role of digital genres and their impact on society (e.g. increasing citizens' literacy in the humanities and social sciences fields).

The second issue, as we observed in Chapter 3, is that even though digital genres play an important role in facilitating researchers' activity and making research processes and outcomes more accessible to diverse publics, it is necessary to consider critically whether they actually democratize research and, if they do, to what extent and how. Future research would need to empirically investigate citizens' responses to and engagement in those genres.

The third issue related to limitations concerns power and access. As we have discussed, whereas digital technologies afford fast access, tensions and conflicting interests exist that may affect the current model of publishing and knowledge circulation, which are, therefore, worth researching. The various stakeholders involved in the production and communication of knowledge online (e.g. researchers, publishing companies, for-profit ASNSs such as Academia.edu, public institutions and research centers) have often-conflicting interests. Thus, it is worth studying what these interests are, and what measures are being taken to defend them. More research is needed on the limitations that major players such as Elsevier, Google and Academia.edu appear to place on the affordances of digital technologies in terms of access, searchability and circulation of knowledge to advance their (competing) financial interests. We do not know how the digitization of academic knowledge and genres may be disrupted or co-opted by prevailing regimes of knowledge production and circulation at local, institutional, national and international levels. Understanding how knowledge is communicated online requires new research on the underlying economics of academic knowledge production and communication in the era of Web 2.0. If digital infrastructures and forms of scholarly communication are determined by economic agendas, we will likely observe changes in scholars' patterns of knowledge production and dissemination that may not always be scholars' preferred modes of social interaction. These are all critical areas for further investigation.

References

Aalbersberg, I.J., Atzeni, S., Koers, H., Specker, B. and Zudilova-Seinstra, E. (2014) Bringing digital science deep inside the scientific article: The Elsevier Article of the Future project. *Liber Quarterly* 23 (4), 274–299.

Acord, S. and Harley, D. (2012) Credit, time and personality: The human challenges to sharing scholarly. *New Media & Society* 15, 379–397.

Adami, E. (2014) Retweeting, reposting, repining, reshaping identities online: Towards a social semiotic multimodal analysis of digital remediation. *Lingue e letterature d'Oriente e d'Occidente* 3, 223–243.

Ädel, A. (2011) Rapport building in student group work. *Journal of Pragmatics* 43, 2932–2947.

Akers, K.G. (2018) F1000Prime: Expert recommendations of journal articles in biology and medicine. *Issues in Science and Technology Libraries* 90. https://doi.org/10.5062/F4Q52MVR (accessed 20 April 2021).

Albornoz, D., Hillyer, B., Posada, A., Okune, A. and Chan, L. (2019) Principles for an inclusive open science: The OCSDNet manifesto. In L. Chan, A. Okune, R. Hillyer, D. Albornoz and A. Posada (eds) *Contextualizing Openness: Situating Open Science* (pp. 23–52). Ottawa: University of Ottawa Press.

Allgaier, J. (2018) Science and medicine on YouTube. In J. Hunsinger, L. Klastrup, M.M. Allen and M. Mathew (eds) *Second International Handbook of Internet Research* (pp. 1–21). Wiesbaden: Springer.

Alperin, J.P., Fischman, G.E. and Willinsky, J. (2008) Open access and scholarly publishing in Latin America: Ten flavours and a few reflections. *Liinc em Revista* 4 (2), 172–185.

Androutsopoulos, J. (2007) Language choice and code-switching in German-based diasporic web forums. In B. Danet and S.C. Herring (eds) *The Multilingual Internet: Language, Culture, and Communication Online* (pp. 340–361). Oxford: Oxford University Press.

Androutsopoulos, J. (2014) Languaging when contexts collapse: Audience design in social networking. *Discourse, Context & Media* 4–5, 62–73.

Androutsopoulos, J. (2015) Networked multilingualism: Some language practices on Facebook and their implications. *International Journal of Bilingualism* 19 (2), 185–205.

Askehave, I. and Swales, J.M. (2001) Genre identification and communicative purpose: A problem and a possible solution. *Applied Linguistics* 21, 195–212.

Askehave, I. and Nielsen, A.E. (2005) Digital genres: A challenge to traditional genre theory. *Information, Technology & People* 18 (2), 120–141.

Attardo, S. (2001) *Humorous Texts: A Semantic and Pragmatic Analysis*. Berlin: Mouton de Gruyter.

Baker, M. (2016) 1,500 scientists lift the lid on reproducibility. *Nature* 533 (7604), 452–454.

Ball, C.E. (2016) The shifting genres of scholarly multimedia: Webtexts as innovation. *The Journal of Media Innovations* (2), 52–71.

Barata, G., Shores, K. and Alperin, J.P. (2018) Local chatter or international buzz? Language differences on posts about Zika research on Twitter and Facebook. *PloS One* 13 (1), e0190482. doi:10.1371/journal.pone.0190482 (accessed 20 April 2021).

Barbour, K. and Marshall, D. (2012) The academic online: Constructing persona through the World Wide Web. *First Monday*, 17 (9). https://firstmonday.org/ojs/index.php/fm/article/view/3969 (accessed 20 April 2021).

Baron, N. (2010) *Escape from the Ivory Tower.* Washington, DC: Island Press.

Bartling, S. and Friesike, S. (2014) Towards another scientific revolution. In S. Bartling and S. Friesike (eds) *Opening Science* (pp. 3–15). Cham: Springer.

Barton, D. and Lee, C. (2013) *Language Online: Investigating Digital Texts and Practices.* Abingdon: Routledge.

Barton, D. and Lee, C. (2017) Methodologies for researching multilingual online texts and practices. In M. Martin-Jones and D. Martin (eds) *Researching Multilingualism: Critical and Ethnographic Perspectives* (pp. 141–154). Abingdon: Routledge.

Barton, D. and McCulloch, S. (2018) Negotiating tensions around new forms of academic writing. *Discourse, Context & Media* 24, 8–15.

Bateman, J.A. (2008) *Multimodality and Genre.* Basingstoke: Palgrave Macmillan.

Bateman, J.A. (2014) Genre in the age of multimodality: Some conceptual refinements for practical analysis. In P. Evangelisti-Allori, V.K. Bhatia and J.A. Bateman (eds) *Evolution in Genre: Emergence, Variation, Multimodality* (pp. 237–269). Frankfurt am Main: Peter Lang.

Bawarshi, A.S. and Reiff, M.J. (2010) *Genre: An Introduction to History, Theory, Research and Pedagogy.* Anderson, SC: Parlor Press.

Baym, N.K. (2010) *Personal Connections in the Digital Age.* London: Polity.

Bazerman, C. (1994) Systems of genres and the enactment of social intentions. In A. Freedman and P. Medway (eds) *Genre and the New Rhetoric* (pp. 79–101). Bristol, PA: Taylor & Francis.

Bazerman, C. (1997) The life of genre, the life in the classroom. In W. Bishop and H. Ostrom (eds) *Genre and Writing* (pp. 19–26). Portsmouth: Boynton/Cook.

Bazerman, C. (2016) Social changes in science communication: Rattling the information chain. In J. Buehl and A. Gross (eds) *Science and the Internet: Communicating Knowledge in a Digital Age* (pp. 267–282). Amityville, NY: Baywood Publishing.

Bennett, K. (2014) Introduction. The political and economic infrastructure of academic practice: The 'semiperiphery' as a category for social and linguistic analysis. In K. Bennett (ed.) *The Semiperiphery of Academic Writing* (pp. 1–9). Basingstoke: Palgrave Macmillan.

Berkenkotter, C. and Huckin, T.N. (1995) *Genre Knowledge in Disciplinary Communication Cognition/Culture/Power.* Hillsdale, NJ: Lawrence Erlbaum.

Berkowitz, J. (2013) Video abstracts, the latest trend in scientific publishing: Will 'publish or perish' soon include 'video or vanish'? *University Affairs.* See http://www.universityaffairs.ca/features/feature-article/video-abstracts-thelatest-trend-in-scientific-publishing/ (accessed 20 April 2021).

Bernstein, R. (2015) Barring it all. *Science.* See https://www.sciencemag.org/careers/2015/01/baring-it-all (accessed 24 July 2021).

Bex, R.T., Lundgren, L. and Crippen, K.J. (2019) Scientific Twitter: The flow of paleontological communication across a topic network. *PloS One* 14 (7): e0219688. doi:10.1371/journal.pone.0219688 (accessed 20 April 2021).

Bhatia, V.K. (2002) Applied genre analysis: A multi-perspective model. *Ibérica. Journal of the European Association of Languages for Specific Purposes* 4, 3–19.

Bhatia, V.K. (2004) *Worlds of Written Discourse: A Genre-Based View.* London: Continuum International.

Biber, D., Connor, U. and Upton, T.A. (2007) *Discourse on the Move: Using Corpus Analysis to Describe Discourse Structure*. Amsterdam: John Benjamins.

Bik, H.M. and Goldstein, M.C. (2013) An introduction to social media for scientists. *PLoS Biology* 11 (4), e1001535. https://doi.org/10.1371/journal.pbio.1001535 (accessed 20 April 2021).

Bjerglund, A.N. and Söderqvist, T. (2012) *Social Media and Public Health Research. Working Paper/Technical Report*. University of Copenhagen. See https://bjerglund.files.wordpress.com/2012/11/final-social-media-and-public-health-research1.pdf (accessed 20 April 2021).

Bloomfield, E.F. and Tillery, D. (2019) The circulation of climate change denial online: Rhetorical and networking strategies on Facebook. *Environmental Communication* 13 (2), 23–34.

Bolter, J.D. and Grusin, R. (1999) *Remediation: Understanding New Media*. Cambridge, MA: The MIT Press.

Bombaci, S.P., Farr, C.M., Gallo, H.T., Mangan, A.M., Stinson. L.T., Kaushik. M. and Pejchar, L. (2016) Using Twitter to communicate conservation science from a professional conference. *Conservation Biology* 30, 216–225.

Bonaiuti, G. (2015) Academic social networks: How the web is changing our way to make and communicate researches. *Research on Education and Media* 7 (2), 3–14.

Bond, S. (2017) Dear scholars, delete your account at Academia.Edu. *Forbes*. See https://www.forbes.com/sites/drsarahbond/2017/01/23/dear-scholars-delete-your-account-at-academia-edu/#7db9e6e92d62 (accessed 18 May 2021).

Bondi, M. (2018) Blogs as interwoven polylogues. The dialogic action game. *Language and Dialogue* 8 (1), 43–65.

Bonney, R., Ballard, H.L., Jordan, R., McCallie, E., Phillips, T.B., Shirk, J. and Wilderman, C.C. (2009) *Public Participation in Scientific Research: Defining the Field and Assessing Its Potential for Informal Science Education*. A CAISE Inquiry Group Report. See https://files.eric.ed.gov/fulltext/ED519688.pdf (accessed 20 April 2021).

Bonney, R., Phillips, T.B., Ballard, H.L. and Enck, J.W. (2016) Can citizen science enhance public understanding of science? *Public Understanding of Science* 25 (1), 2–16.

Bornmann, L., Wolf, M. and Daniel, M.D. (2012) Closed versus open reviewing of journal manuscripts: How far do comments differ in language use? *Scientometrics* 91 (3), 843–856.

Boulton, G. (2012) *Science as an Open Enterprise*. London: The Royal Society. See http://royalsociety.org/policy/projects/science-public-enterprise/report/. (accessed 20 April 2021).

Bowker, L. (2002) *Computer Aided Translation Technology: A Practical Introduction*. Ottawa: University of Ottawa Press.

boyd, d.m. (2002) Faceted identity: Managing representation in a digital world. Unpublished master's thesis, Massachusetts Institute of Technology.

boyd, d.m. (2010) Social network sites as networked publics: Affordances, dynamics, and implications. In Z. Papacharissi (ed.) *A Networked Self: Identity, Community and Culture on Social Network Sites* (pp. 39–58). New York: Routledge.

boyd, d.m. and Ellison, N.B. (2007) Social network sites: Definition, history, and scholarship. *Journal of Computer-Mediated Communication* 13 (1), 210–230.

Boyd, S. (2003) Are you ready for social software? *Darwin Magazine (IDG)*. See https://stoweboyd.com/post/2325281845/are-you-ready-for-social-software (accessed 25 July 2021).

Bradley, J.C. (2006) Open Notebook Science. Blog post on *Drexel COAS E-learning Blog*. 26 September. See http://drexel-coas-elearning.blogspot.com.br/2006/09/open-notebook-science.html (accessed 20 March 2020).

Bradley, J.C. (2007) C&E news article on writing. Blog post on UsefulChem (Blog). 13 August. http://usefulchem.blogspot.com.br/2007/08/c-news-article-onwriting.html (accessed 20 March 2020).

Breeze, R. (2019) Continuity and change: Negotiating relationships in traditional and online peer review genres. In M.J. Luzón and C. Pérez-Llantada (eds) *Science Communication on the Internet: Old Genres Meet New Genres* (pp. 107–129). Amsterdam: John Benjamins.

Breure, L., Voorbij, H. and Hoogerwerf, M. (2011) Rich Internet publications: 'Show what you tell'. *Journal of Digital Information* 12 (1). https://journals.tdl.org/jodi/index.php/jodi/article/view/1606/1738 (accessed 20 April 2021).

Brossard, D. (2013) New media landscapes and the science information consumer. *Proceedings of the National Academy of Sciences of the United States of America* 110 (Suppl. 3), 14096–14101.

Brossard, D. and Scheufele, D.A. (2013) Science, new media, and the public. *Science* 339 (6115), 40–41.

Brunner, M. (2016) Dissemination 2.0: Bench, bedside, and beyond. *Substance Abuse Library and Information Studies: Proceedings of the 38th Annual SALIS/AMHL Conference*. See http://salis.org/salisjournal/vol3/25_Brunner_2016.pdf (accessed 20 April 2021).

Bucholtz, M. and Hall, K. (2005) Identity and interaction: A sociocultural linguistic approach. *Discourse Studies* 7 (4–5), 585–614.

Budden, A.E., Tregenza, T., Aarssen, L.W., Koricheva, J., Leimu, R. and Lortie C.J. (2008) Double-blind review favours increased representation of female authors. *Trends in Ecology and Evolution* 23 (1), 4–6.

Budge, K., Lemon, N. and McPherson, M. (2016) Academics who tweet: 'Messy' identities in academia. *Journal of Applied Research in Higher Education* 8 (2), 210–221.

Buehl, J. (2016) Revolution or evolution? Casing the impact of digital media on the rhetoric of science. In A. Gross and J. Buehl (eds) *Science and the Internet: Communicating Knowledge in a Digital Age* (pp. 1–9). Amityville, NY: Baywood Publishing.

Bukvova, H. (2011) Scientists online: A framework for the analysis of internet profiles. *First Monday* 16, 10. See http://firstmonday.org/ojs/index.php/fm/article/view/3584/3065. (accessed 20 April 2021).

Caliendo, G. (2014) The popularization of science in web-based genres. In G. Caliendo and G. Bongo (eds) *The Language of Popularization: Theoretical and Descriptive Models* (pp. 101–132). Bern: Peter Lang.

Calsamiglia, H. and Van Dijk, T.A. (2004) Popularization discourse and knowledge about the genome. *Discourse & Society* 15 (4), 369–389.

Canagarajah, A.S. (1996) 'Nondiscursive' requirements in academic publishing, material resources of periphery scholars, and the politics of knowledge production. *Written Communication* 13, 435–472.

Canagarajah, A.S. (2002) *A Geopolitics of Academic Writing*. Pittsburgh, PA: University of Pittsburgh Press.

Canagarajah, A.S. (2018) Translingual practice as spatial repertoires: Expanding the paradigm beyond structuralist orientations. *Applied Linguistics* 39 (1), 31–54.

Carter-Thomas, S. and Rowley-Jolivet, E. (2017) Open science notebooks: New insights, new affordances. *Journal of Pragmatics* 116, 64–76.

Casper, C. (2009) Mutable mobiles: Online journals and the evolving genre ecosystem of science. PhD thesis, North Carolina State University. See https://repository.lib.ncsu.edu/handle/1840.16/3905 (accessed 20 April 2021).

Casper, C. (2016) The online research article and the ecological basis of new digital genres. In A. Gross and J. Buehl (eds) *Science and the Internet: Communicating Knowledge in a Digital Age* (pp. 77–98). London: Routledge.

Catenaccio, P. (2012) A genre-theory approach to the website: Some preliminary considerations. In S. Campagna, G. Garzone, C. Ilie and E. Rowley-Jolivet (eds) *Evolving Genres in Web Mediated Communication* (pp. 27–52). Bern: Peter Lang.

Cavalieri, S. (2020) Broadcasting medical discourse: The dissemination of dietary treatments for refractory epilepsy through YouTube. In S. Maci and M. Sala (eds) *Scholarly Pathways: Knowledge Transfer and Knowledge Exchange in Academia* (pp. 433–450). Bern: Peter Lang.

Chapman, A. and Greenhow, C. (2019) Citizen-scholars: Social media and the changing nature of scholarship. *Publications* 7 (1), 11. https://doi.org/10.3390/publications7010011 (accessed 20 April 2021).

Chen, H. (2013) Identity practices of multilingual writers in social networking spaces. *Language Learning & Technology* 17 (2), 143–170.

Chivite, I. (2017) Using Survey123 for citizen science and crowdsourcing initiatives. 2017 Education GIS Conference Proceedings. See https://proceedings.esri.com/library/userconf/educ17/papers/educ_99.pdf (accessed 17 March 2020).

Clinio, A. and Albagli, S. (2017) Open Notebook Science as an emerging epistemic culture within the Open Science Movement. *Revue Française des Sciences de L'information et de La Communication* 11. doi:10.4000/rfsic.3186 (accessed 20 April 2021).

Cody, E., Reagan, A., Mitchell, L., Dodds, P.S. and Danforth, C. (2015) Climate change sentiment on Twitter: An unsolicited public opinion poll. *PLoS One* 10 (8), e0136092. doi:10.1371/journal.pone.0136092 (accessed 20 April 2021).

Collins, K., Shiffman, D. and Rock, J. (2016) How are scientists using social media in the workplace? *PLoS One* 11 (10), e0162680. doi:10.1371/journal.pone.0162680 (accessed 20 April 2021).

Cope, B. and Kalantzis, M. (2014) Changing knowledge ecologies and the transformation of the scholarly journal. In B. Cope and A. Phillips (eds) *The Future of the Academic Journal* (pp. 9–83). Oxford: Chandos Publishing.

Corcoran, J.N., Englander, K. and Muresan, L. (2019) Diverse global perspectives on scholarly writing for publication. In K. Englander and L.M. Muresan (eds) *Pedagogies and Policies on Publishing Research in English: Local Initiatives Supporting International Scholars* (pp. 1–16). New York: Routledge.

Crane, D. (1969) Social structure in a group of scientists: A test of the invisible college hypothesis. *American Sociological Review* 34, 335–352.

Cribb, J. and Hartomo, T.S. (2010) *Open Science: Sharing Knowledge in the Global Century*. Melbourne: CSIRO Pub.

Crowston, K. and Williams, M. (2000) Reproduced and emergent genres of communication on the World Wide Web. *The Information Society* 16 (3), 201–215.

Culpeper, J. (1996) Towards an anatomy of impoliteness. *Journal of Pragmatics* 25, 349–367.

Curry, M.J. and Lillis, T.M. (2010) Academic research networks: Accessing resources for English-medium publishing. *English for Specific Purposes* 29 (4), 281–295.

Curry, M.J. and Lillis, T. (2018) Problematizing English as the privileged language of global academic publishing. In M.J. Curry and T. Lillis (eds) *Global Academic Publishing: Policies, Perspectives and Pedagogies* (pp. 1–20). Bristol: Multilingual Matters.

Dallmeier-Tiessen, S. and Simko, T. (2019) Open Science: A vision of collaborative reproducible and reusable research. *CERN Courier*. See https://cerncourier.com/a/open-science-a-vision-for-collaborative-reproducible-and-reusable-research/ (accessed 20 April 2021).

Danet, B. and Herring, S.C. (eds) (2007) *The Multilingual Internet: Language, Culture, and Communication Online*. Oxford: Oxford University Press.

Darling, E.S., Shiffman, D., Côté, I.M. and Drew, J.A. (2013) The role of Twitter in the life cycle of a scientific publication. *Ideas in Ecology and Evolution* 6 (1), 32–43.

Darvin, R. (2016) Language and identity in the digital age. In S. Preece (ed.) *Routledge Handbook of Language and Identity* (pp. 523–540). Abingdon: Routledge.

Davies, B. and Harré, R. (1990) Positioning: The discursive production of selves. *Journal for the Theory of Social Behaviour* 20, 43–63.

Davies, J. and Merchant, G. (2007) Looking from the inside out: Academic blogging as new literacy. In M. Knobel and C. Lankshear (eds) *A New Literacies Sampler* (pp. 167–197). New York: Peter Lang.

Dawson, D. (2012) Open science and crowd science: Selected sites and resources. *Issues in Science and Technology Librarianship* 69. See https://scholar.uwindsor.ca/lripub/44. (accessed 20 April 2021).

De Groot, E., Nickerson, C., Korzilius, H. and Gerritsen, M. (2016) Picture this: Developing a model for the analysis of visual metadiscourse. *Journal of Business and Technical Communication* 30 (2), 165–201.

Delfanti, A. (2008) Collaborative web between open and closed science. *Journal of Science Communication* (JCOM) 7. See https://jcom.sissa.it/sites/default/files/documents/Jcom0702(2008)C01.pdf (accessed 20 April 2021).

Demarest, B. and Sugimoto, C.S. (2014) Argue, observe, assess: Measuring disciplinary identities and differences through socio-epistemic discourse. *Journal of the Association for Information Science and Technology* 66 (7), 1374–1387.

Demeter, M. and Istratii, R. (2020) Scrutinising what open access journals mean for global inequalities. *Publishing Research Quarterly* 36, 505–522.

Deng, S., Tong, J. and Fu, S. (2018) Interaction on an academic social networking sites: A study of ResearchGate Q&A on Library and Information Science. In *JCDL '18: Proceedings of the 18th ACM/IEEE on Joint Conference on Digital Libraries* (pp. 25–28). New York: ACM Press. doi:10.1145/3197026.3197062

Dennen, V.P. (2009) Constructing academic alter-egos: Identity issues in a blog-based community. *Identity in the Information Society* 2 (1), 23–38.

Devitt, A.J. (2004) *Writing Genres*. Carbondale: Southern Illinois University Press.

Devitt, A.J. (2009) Re-fusing form in genre study. In J. Giltrow and D. Stein (eds) *Genres in the Internet: Issues in the Theory of Genre* (pp. 27–47). Amsterdam: John Benjamins.

Devitt, A.J. (2015) Genre performances: John Swales' genre analysis and rhetorical-linguistic genre studies. *Journal of English for Academic Purposes* 19, 44–51.

Didegah, F., Mejlgaard, N. and Sørensen, M. (2018) Investigating the quality of interactions and public engagement around scientific papers on Twitter. *Journal of Informetrics* 12, 960–971. doi: 10.1016/j.joi.2018.08.002.

Donath, J. and boyd, d.m. (2004) Public displays of connection. *BT Technology Journal* 22, 71–82.

Dudley-Evans, A. (2000) Genre analysis: A key to a theory of ESP? *Ibérica* 2, 3–11.

Dumont, K. and Frindte, W. (2005) Content analysis of the homepages of academic psychologists. *Computers in Human Behavior* 21, 73–83.

Duszak, A. and Lewkowicz, J. (2008) Publishing academic texts in English: A Polish perspective. *Journal of English for Academic Purposes* 7 (2), 108–120.

eLife (2020) The review process. See https://reviewer.elifesciences.org/reviewer-guide/review-process (accessed 20 April 2021).

Elsayed, A.M. (2016) The use of academic social networks among Arab researchers: A survey. *Social Science Computer Review* 34 (3), 378–391.

EMBO Journal (2020) Author guidelines. See https://www.embopress.org/page/journal/14602075/authorguide#transparentprocess (accessed 20 April 2021).

Enago Academy (2018) Exploring video journals: An interview with latest thinking. See https://www.enago.com/academy/exploring-video-journals-an-interview-with-latest-thinking/ (accessed 20 April 2021).

Engberg, J. and Maier, C.D. (2015) Exploring the hypermodal communication of academic knowledge beyond generic structure. In M. Bondi, S. Cacchiani and D. Mazzi (eds) *Discourse in and through the Media: Recontextualizing and Reconceptualizing Expert Discourse* (pp. 46–62). Newcastle upon Tyne: Cambridge Scholars Publishing.

Engberg, J. and Maier, C.D. (2019) Researchers' move from page to screen: Addressing the effects of the video article format upon academic user engagement and knowledge-building processes. In C.S. Guinda (ed.) *Engagement in Professional Genres* (pp. 179–195). Amsterdam: John Benjamins.

Erviti, M.C. and Stengler, E. (2016) Online science videos: An exploratory study with major professional content providers in the United Kingdom. *Journal of Science Communication* 15 (6), A06. http://eprints.uwe.ac.uk/30236 (accessed 20 April 2021).

European Commission (2015) Study on Open Science. Impact, implications and policy options. *European Commission, Research and Innovation.* See https://op.europa.eu/en/publication-detail/-/publication/b0906f2d-8907-11e5-b8b7-01aa75ed71a1 (accessed 20 April 2021).

Faculty Opinions (2020) https://f1000.com/prime/home. (accessed 20 April 2021).

F1000Research (2020) Submit your research. See https://f1000research.com/for-authors/publish-your-research (accessed 20 April 2021).

Fahnestock, J. (1986) Accommodating science. The rhetorical life of scientific facts. *Written Communication* 3 (3), 275–296.

Fahnestock, J. (2016) Controversies on the web: The case of adult human neurogenesis. In A. Gross and J. Buehl (eds) *Science and the Internet: Communicating Knowledge in a Digital Age* (pp. 115–141). Amityville, NY: Baywood Publishing.

Fazel, I. and Heng Hartse, J. (2018) Reconsidering 'predatory' open access journals in an age of globalised English language academic publishing. In M.J. Curry and T. Lillis (eds) *Global Academic Publishing: Policies, Perspectives and Pedagogies* (pp. 200–213). Bristol: Multilingual Matters.

Fecher, B. and Friesike, S. (2014) Open science: One term, five schools of thought. In S. Bartling and S. Friesike (eds) *Opening Science* (pp. 17–47). Cham: Springer.

Ferguson, G., Pérez-Llantada, C. and Plo, R. (2011) English as an international language of scientific publication: A study of attitudes. *World Englishes* 30, 41–59.

Finkler, W. and León, B. (2019) The power of storytelling and video: A visual rhetoric for science communication. *JCOM 18* (05), A02. https://doi.org/10.22323/2.18050202. (accessed 20 April 2021).

Firth, A. (1996) The discursive accomplishment of normality: On conversation analysis and 'lingua franca' English. *Journal of Pragmatics* 26 (2), 237–259.

Flowerdew, J. and Wang, S.H. (2015) Identity in academic discourse. *Annual Review of Applied Linguistics* 35, 81–99.

Flowerdew, J. and Wang, S.H. (2016) Author's editor revisions to manuscripts published in international journals. *Journal of Second Language Writing* 32, 39–52.

Flowerdew, L. (2012) *Corpora and Language Education.* London: Palgrave Macmillan.

Follett, R. and Strezov, V. (2015) An analysis of citizen science based research: Usage and publication patterns. *PLoS One* 10 (11), e0143687. https://doi.org/10.1371/journal.pone.0143687 (accessed 20 April 2021).

Freadman, A. (2002) Uptake. In R.M. Coe, L. Lingard and T. Teslenko (eds) *The Rhetoric and Ideology of Genre: Strategies for Stability and Change* (pp. 39–53). Cresskill, NJ: Hampton Press.

Freadman, A. (2012) The traps and trappings of genre theory. *Applied Linguistics* 33 (5), 544–563.

Frontiers (2016) About Frontiers academic journals and research community. See https://www.frontiersin.org/about/about-frontiers (accessed 20 April 2021).

Frontiers (2020) Collaborative peer review. See https://www.frontiersin.org/about/review-system (accessed 20 April 2021).

Gardiner, A., Sullivan, M. and Grand, A. (2018) Who are you writing for? Differences in response to blog design between scientists and nonscientists. *Science Communication* 40 (1), 109–123.

Gee, J.P. (2004) *Situated Language and Learning: A Critique of Traditional Schooling*. New York: Routledge.

Gibson, J.J. (1966) *The Senses Considered as Perceptual Systems*. Boston, MA: Houghton Mifflin.

Gilbert, G.N. and Mulkay, M. (1984) *Opening Pandora's Box: A Sociological Analysis of Scientists' Discourse*. Cambridge: Cambridge University Press.

Giltrow, J. and Stein, D. (eds) (2009) *Genres in the Internet. Issues in the Theory of Genre*. Amsterdam: John Benjamins.

Goffman, E. (1959) *The Presentation of Self in Everyday Life*. New York: Doubleday.

Goodwin, S., Jeng, W. and He, D. (2014) Changing communication on ResearchGate through interface updates. *Proceedings of the American Society for Information Science and Technology* 51 (1), 1–4.

Görögh, E., Ross-Hellauer, T., Schmidt, B., Bardi, A., Casarosa, V., Manghi, P., Zendel, O., Luzi, D. and Ruggieri, R. (2019) Deliverable D3.4: Open Peer Review: Good practices and lessons learned. Zenodo. See https://zenodo.org/record/2557308#.XqnBsKgzbIU (accessed 20 April 2021).

Gotti, M. (2014) Reformulation and recontextualisation in popularisation discourse. *Ibérica. Journal of the European Association of Languages for Specific Purposes* 27, 15–34.

Graupner, M. (2010) Blogging a scholarly identity. *Computers and Composition Online*. Spring 2010. See http://cconlinejournal.org/graupner2010/Graupner_Blogging.pdf. (accessed 20 April 2021).

Gross, A.G. and Harmon, J. (2016) *The Internet Revolution in the Sciences and Humanities*. New York: Oxford University Press.

Gruzd, A., Staves, K. and Wilk, A. (2012) Connected scholars: Examining the role of social media in research practices of faculty using the UTAUT model. *Computers in Human Behavior* 28 (6), 2340–2350.

Hafner, C.A. (2018) Genre innovation and multimodal expression in scholarly communication: Video methods articles in experimental biology. *Ibérica, Journal of the European Association of Languages for Specific Purposes* 36, 15–41.

Hafner, C.A. and Ho, W.Y.J. (2020) Assessing digital multimodal composing in second language writing: Towards a process-based model. *Journal of Second Language Writing* 47, 100710.

Hall, G. (2015) What does Academia.edu's success mean for Open Access?. *LSE blog*. See https://blogs.lse.ac.uk/impactofsocialsciences/2015/10/22/does-academia-edu-mean-open-access-is-becoming-irrelevant/ (accessed 18 April 2021).

Hampton, S.E., Anderson, S.S., Bagby, S.C., Gries, C., Han, X., Hart, E.M., Jones, M.B., Lenhardt, W.C., MacDonald, A., Michener, W.K., Mudge, J., Pourmokgtarian, A., Schildhauer, M.P., Woo, K.H. and Zimmerman, N. (2015) The Tao of Open Science for ecology. *Ecosphere* 6 (7), 1–13.

Han, Y. (2019) Translanguaging as transnational spaces: Chinese visiting scholars' language practices on WeChat. *International Journal of Multilingualism* 1 (22), 174–195. doi:10.1080/14790718.2018.1546308

Hanauer, D.I., Sheridan, C.L. and Englander, K. (2019) Linguistic injustice in the writing of research articles in English as a second language: Data from Taiwanese and Mexican researchers. *Written Communication* 36, 136–154.

Hara, N., Abbazio, J. and Perkins, K. (2019) An emerging form of public engagement with science: Ask Me Anything (AMA) sessions on Reddit r/ science. *PLoS One* 14 (5), e0216789. https://doi.org/10.1371/journal.pone.0216789 (accessed 20 April 2021).

Harmon, J. (2019) At the frontiers of the online scientific article. In M.J. Luzón and C. Pérez-Llantada (eds) *Science Communication on the Internet: Old Genres Meet New Genres* (pp. 19–40). Amsterdam: John Benjamins.

Heap, T. and Minocha, S. (2012) An empirically grounded framework to guide blogging for digital scholarship. *Research in Learning Technology* 20, 176–188.
Henderson, K. (2013) Reinventing the methods journal: Increasing reproducibility with video journals. *Against the Grain* 25 (5), Article 9. https://doi.org/10.7771/2380-176X.6605 (accessed 20 April 2021).
Hendges, G. and Florek, C. (2019) The graphical abstract as a new genre in the promotion of science. In M.J. Luzón and C. Pérez-Llantada (eds) *Science Communication on the Internet: Old Genres Meet New Genres* (pp. 59–80). Amsterdam: John Benjamins.
Herman, E. (2018) Scholarly reputation. *FEMS Microbiology Letters* 365 (18), fny200. https://doi.org/10.1093/femsle/fny200 (accessed 20 April 2021).
Herman, E. and Nicholas, D. (2019) Scholarly reputation building in the digital age: An activity-specific approach. *El Profesional de la Información* 28 (1), e280102. https://doi.org/10.3145/epi.2019.ene.02 (accessed 20 April 2021).
Herring, S.C. (2013) Discourse in Web 2.0: Familiar, reconfigured, and emergent. In D. Tannen (ed.) *Discourse 2.0. Language and New Media* (pp. 1–255). Washington DC: Georgetown University Press.
Herring, S.C. (2019) The co-evolution of computer-mediated communication and computer-mediated discourse analysis. In P. Bou-Franch and P. Garcés-Conejos Blitvich (eds) *Analysing Digital Discourse: New insights and Future Directions* (pp. 25–67). London: Palgrave Macmillan.
Herring, S., Scheidt, L.A., Bonus, S. and Wright, E. (2004) Bridging the gap: A genre analysis of weblogs. In *Proceedings of the 37th Hawaii International Conference on System Sciences* (HICSS-37) (pp. 1–11). Los Alamitos: IEEE Computer Society Press.
Heyd, T. (2008) *Email Hoaxes: Form, Function, Genre Ecology*. Amsterdam: John Benjamins.
Heyd, T. (2016) Digital genres and processes of remediation. In A. Georgakopoulou and T. Spilioti (eds) *The Routledge Handbook of Language and Digital Communication* (pp. 87–102). London: Routledge.
Hildebrandt, K. and Couros, A. (2016) Digital selves, digital scholars: Theorising academic identity in online spaces. *Journal of Applied Social Theory* 1 (1), 87–100.
Hine, C. (2002) Cyberscience and social boundaries: The implications of laboratory talk on the Internet. *Sociological Research Online* 7 (2). See http://www.socresonline.org.uk/7/2/hine.html (accessed 20 April 2021).
Hoeberechts, D., Owens, Riddell, D.J. and Robertson, A.D. (2015) The power of seeing: Experiences using video as a deep-sea engagement and education tool. *MTS/IEEE Oceans* (pp. 1–9). Washington: IEEE.
Hopkins, J. (2015) Assembling blog affordances: Theorising affordances and agency in new media. *Working Paper for the EASA Media Anthropology Network's 51st e-Seminar*. See https://www.easaonline.org/downloads/networks/media/51p.pdf (accessed 20 April 2021).
Hormia-Poutanen, K. and Forsström, P. (2016) Collaboration at international, national and institutional level: Vital in fostering Open Science. *LIBER Quarterly* 26 (1), 3–12.
Hudson, R.C., Duncan, S. and Reeve, C. (2015) Affinity spaces for informal science learning: Developing a research agenda. See http://www.informalscience.org/sites/default/files/AffinitySpacesFinalReport.pdf (accessed 20 April 2021).
Huggett, J. (2019) Resilient scholarship in the digital age. *Journal of Computer Applications in Archaeology* 2 (1), 105–119.
Hunston, S. (2005) Conflict and consensus: Construing opposition in applied linguistics. In E. Tognini-Bonelli and G.D. Camiciotti (eds) *Strategies in Academic Discourse* (pp. 1–15). Philadelphia, PA: John Benjamins.
Hutchby, I. (2001) Technologies, texts and affordances. *Sociology* 35 (2), 441–456.
Hyland, K. (2000) *Disciplinary Discourses: Social Interactions in Academic Writing*. London: Pearson Education Limited.

Hyland, K. (2005) Stance and engagement: A model of interaction in academic discourse. *Discourse Studies* 7, 173–192.
Hyland, K. (2010) Constructing proximity: Relating to readers in popular and professional science. *Journal of English for Academic Purposes* 9, 116–127.
Hyland, K. (2011) The presentation of self in scholarly life: Identity and marginalization in academic homepages. *English for Specific Purposes* 30 (4), 286–297.
Hyland, K. (2012) Individuality or conformity? Identity in personal and university academic homepages. *Computers and Composition* 29 (4), 309–322.
Hyland, K. (2015a) Genre, discipline and identity. *Journal of English for Academic Purposes* 19, 32–43.
Hyland, K. (2015b) *Teaching and Researching Writing*. New York: Routledge.
Hyland, K. and Jiang, K. (2017) Is academic writing becoming more informal? *English for Specific Purposes* 45, 40–51.
Hynninen, N. and Kuteeva, M. (2017) 'Good' and 'acceptable' English in L2 research writing: Ideals and realities in history and computer science. *Journal of English for Academic Purposes* 30, 53–65.
Internet World Statistics (2019) See http://www.internetworldstatistics.com/
Jahng, M.R. and Lee, N. (2018) When scientists tweet for social changes: Dialogic communication and collective mobilization strategies by Flint Water Study scientists on Twitter. *Science Communication* 40 (1), 89–108.
Jakubowicz, A. and van Leeuwen, T. (2010) The Goldberg Variations 1: Assessing the quality of multidimensional linear texts and their re-emergence in multimedia publications. *Discourse and Communication* 4 (4), 361–378.
Jarreau, P.B. (2015) All the Science that is Fit to Blog: An Analysis of Science Blogging Practices (doctoral dissertation). Louisiana State University. See https://digitalcommons.lsu.edu/gradschool_dissertations/1051/ (accessed 20 April 2021).
Jeng, W., He, D. and Jiang, J. (2015) User participation in an academic social networking service: A survey of open group users on Mendeley. *Journal of the Association for Information Science and Technology* 66, 890–904.
Jeng, W., Goodwin, S., He, D. and Li, L. (2017) Information exchange on an academic social networking site: A multidiscipline comparison on ResearchGate Q&A. *Journal of the Association for Information Science and Technology* 68 (3), 638–652.
Jewitt, C. (2016) Multimodal analysis. In A. Georgakopoulou and T. Spilioti (eds) The *Routledge Handbook of Language and Digital Communication* (pp. 69–84). London: Routledge.
Jewitt, C., Bezemer, J. and O'Halloran, K. (2016) *Introducing Multimodality*. New York: Routledge.
Jones, R.H. and Hafner, C.A. (2012) *Understanding Digital Literacies: A Practical Introduction*. London: Routledge.
Jordan, K. (2014) Academics and their online networks: Exploring the role of academic social networking sites. *First Monday* 19 (3). http://doi.org/10.5210/fm.v19i11.4937. (accessed 20 April 2021).
Jordan, K. (2017) Understanding the structure and role of academics' ego-networks on social networking sites. PhD thesis, The Open University. https://doi.org/10.21954/ou.ro.0000bc83 (accessed 20 April 2021).
Jordan, K. (2019) From social networks to publishing platforms: A review of the history and scholarship of academic social network sites. *Frontiers in Education* 6, article 5. https://doi.org/10.3389/fdigh.2019.00005 (accessed 20 April 2021).
Kahle, K., Sharon, A.J. and Baram-Tsabari, A. (2016) Footprints of fascination: Digital traces of public engagement with particle physics on CERN's media platforms. *PLoS One* 11 (5), e0156409. https://doi.org/10.1371/journal.pone.0156409 (accessed 20 April 2021).

Kaplan, A.M. and Haenlein, M. (2011) The early bird catches the news: Nine things you should know about micro-blogging. *Business Horizons* 54 (2), 105–113.

Kedrowicz, A.A. and Sullivan, K.R. (2012) Professional identity on the Web: Engineering blogs and self-presentation. *Engineering Studies* 4, 33–53.

Kelland, K. (2020) Speed science: The risks of swiftly spreading coronavirus research. See https://www.reuters.com/article/us-china-health-research-analysis/speed-science-the-risks-of-swiftly-spreading-coronavirus-research-idUSKBN20D21S (accessed 20 March 2020).

Kelly, A.R. and Kittle Autry, M. (2013) Access, accommodation, and science: Knowledge in an 'open' world. *First Monday* 18 (6). See http://firstmonday.org/ojs/index.php/fm/article/view/4341/3684 (accessed 20 April 2021).

Kelly, A.R. and Miller, C. (2016) Intersections: Scientific and para-scientific communication on the Internet. In A. Gross and J. Buehl (eds) *Science and the Internet: Communicating Knowledge in a Digital Age* (pp. 221–245). Amityville, NY: Baywood Publishing.

Kimmons, R. and Veletsianos, G. (2014) The fragmented educator 2.0: Social networking sites. Acceptable identity fragments, and the identity constellation. *Computers & Education* 72, 292–301.

Kirkup, G. (2010) Academic blogging: Academic practice and academic identity. *London Review of Education* 8 (1), 75–84.

Kjellberg, S. (2009) Blogs as interfaces between several worlds: A case study of the Swedish academic blogosphere. *Human IT* 10 (3), 1–45.

Kjellberg, S. (2010) I am a blogging researcher: Motivations for blogging in a scholarly context. *First Monday* 15 (8). http://firstmonday.org/article/view/2962/2580 (accessed 20 April 2021).

Kjellberg, S. (2014) Researchers' blogging practices in two epistemic cultures: The scholarly blog as a situated genre. *Human IT* 12 (3), 36–77.

Knorr-Cetina, K.D. (1981) *The Manufacture of Knowledge. An Essay on the Constructivist and Contextual Nature of Science*. Oxford: Pergamon Press.

Kouper, I. (2010) Science blogs and public engagement with science: Practices, challenges, and opportunities. *Journal of Science Communication* 9 (1), 1–10.

Kourilová, M. (1998) Communicative characteristics of reviews of scientific paper written by non-native users of English. *Endocrine Regulations* 32, 107–114.

Kousha, K., Thelwall, M. and Abdoli, M. (2012) The role of online videos in research communication: A content analysis of YouTube videos cited in academic publications. *Journal of the American Society for Information Science and Technology* 63 (9), 1710–1727.

Kress, G. (2003) *Literacy in the New Media Age*. London: Routledge.

Kress, G. (2010) *Multimodality: A Social Semiotic Approach to Contemporary Communication*. London: Routledge.

Kress, G., Jewitt, C., Ogborn J. and Tsatsarelis C. (2001) *Multimodal Teaching and Learning*. London: Continuum.

Kulavuz-Onal, D. and Vásquez, C. (2018) 'Thanks, shokran, gracias': Translingual practices in a Facebook group. *Language Learning and Technology* 22 (1), 240–255.

Kullenberg, C. and Kasperowski, D. (2016) What is citizen science?: A scientometric meta-analysis. *PLoS One* 11 (1), e0147152. https://doi.org/10.1371/journal.pone.0147152 (accessed 20 April 2021).

Kuteeva, M. and Mauranen, A. (2014) Writing for publication in multilingual contexts: An introduction to the special issue. *Journal of English for Academic Purposes* 13, 1–4.

Kwasnik, B.H. and Crowston, K. (2005) Genre of digital documents. *Introduction to the Special Issue of Information, Technology & People* 18 (2), 76–88.

Kyvik, S. (2005) Popular science publishing and contributions to public discourse among university faculty. *Science Communication* 26 (3), 288–311.

Laakso, M., Lindman, J., Shen, L.C., Nyman, L. and Björk, B.C. (2017) Research output availability on academic social networks: Implications for stakeholders in academic publishing. *Electronic Markets* 27, 125–133.

Lane, S., Karatsolis, A. and Bui, L. (2015) Graphical abstracts: A taxonomy and critique of an emerging genre. In K. Gossett and A. Mallory (eds.) *Proceedings of the 33rd Annual International Conference on the Design of Communication*. New York: ACM. doi: http://dx.doi.org/10.1145/2775441.2775465 (accessed 20 March 2021).

Lee, C. (2016) *Multilingualism Online*. London: Routledge.

Lee, C. and Barton, D. (2011) Constructing glocal identities through multilingual writing practices on Flickr.com. *International Multilingual Research Journal* 5, 1–21.

Lemke, J.L. (1998) Multiplying meaning: Visual and verbal semiotics in scientific text. In J.R. Martin and R. Veel (eds) *Reading Science* (pp. 87–113). London: Routledge.

León, B. and Bourk, M. (2018) Investigating science-related online video. In B. León and M. Bourk (eds) *Communicating Science and Technology through Online Video* (pp. 1–14). New York/London: Routledge.

Leppänen, S. and Peuronen, S. (2012) Multilingualism on the Internet. In M. Martin-Jones, A. Blackledge and A. Creese (eds) *The Routledge Handbook of Multilingualism* (pp. 384–402). London: Routledge.

Li, L., He, D., Zhang, C., Geng, L. and Zhang, K. (2018) Characterizing peer-judged answer quality on academic Q&A sites: A cross-disciplinary case study on ResearchGate. *Aslib Journal of Information Management* 70 (3), 269–287.

Lievrouw, L.A. and Carley, K. (1990) Changing patterns of communication among scientists in an era of 'telescience'. *Technology in Society* 12, 457–477.

Lillis, T.M. and Curry, M.J. (2010) *Academic Writing in a Global Context: The Politics and Practices of Publishing in English*. New York: Routledge.

Lillis, T. and Curry, M.J. (2015) The politics of English, language and uptake: The case of international academic journal article reviews. *AILA Review* 28, 127–150.

Liu, R. (2020) Research video abstracts in the making: A revised move analysis. *Journal of Technical Writing and Communication* 3, 423–446. doi:10.1177/0047281619894981

Lomborg, S. (2011) Social media as communicative genres. *Mediekultur* 27 (51), 55–71.

López-Goñi, I. and Sánchez-Angulo, M. (2018) Social networks as a tool for science communication and public engagement: Focus on Twitter. *FEMS Microbiology Letters* 365 (2), fnx246. https://doi.org/10.1093/femsle/fnx246 (accessed 20 April 2021).

López-Pérez, L. and Olvera-Lobo, M.D. (2016) Social media as channels for the public communication of science. The case of Spanish research centers and public universities. In K. Knautz and K.S. Baran (eds) *Facets of Facebook: Use and Users* (pp. 241–264). Berlin: De Gruyter House.

Lorés-Sanz, R. (2020) Science on the web: The exploration of European research websites of energy-related projects as digital genres for the promotion of values. *Discourse, Context and Media* 35, 1–10. https://doi.org/10.1016/j.dcm.2020.100389

Luce, R. and Di Giacomo, M. (2003) Personalized and collaborative digital library capabilities: Responding to the changing nature of scientific research. *Science & Technology Libraries* 24 (1/2), 135–152.

Lüders, M., Prøitz, L. and Rasmussen, T. (2010) Emerging personal media genres. *New Media & Society* 12 (6), 947–963.

Lupton, D., Mewburn, I. and Thomson, P. (2017) The digital academic: Identities, contexts and politics. In D. Lupton, I. Mewburn and P. Thomson (eds) *The Digital Academic: Critical Perspectives on Digital Technologies in Higher Education* (pp. 1–19). London: Routledge.

Luzón, M.J. (2011) 'Interesting post, but I disagree': Social presence and antisocial behavior in academic weblogs. *Applied Linguistics* 32 (5), 517–540.

Luzón, M.J. (2013) Public communication of science in blogs: Recontextualizing scientific discourse for a diversified audience. *Written Communication* 30, 428–457.

Luzón, M.J. (2014) Engaging in scientific controversies in science blogs: The expression of allegiance and ideological commitment. In H.L. Lim and F. Sudweeks (eds) *Innovative Methods and Technologies for Electronic Discourse Analysis* (pp. 235–249). Hershey, PA: IGI Global.

Luzón, M.J. (2017) Connecting genres and languages in online scholarly communication: An analysis of research group blogs. *Written Communication* 34 (4), 441–471.

Luzón, M.J. (2018a) Constructing academic identities online: Identity performance in research group blogs written by multilingual scholars. *Journal of English for Academic Purposes* 33, 24–39.

Luzón, M.J. (2018b) Features of online ELF in research group blogs written by multilingual scholars. *Discourse, Context & Media* 24, 24–32.

Luzón, M.J. (2018c) Academic social networking sites (ASNS) as ELF settings: An analysis of interactional strategies in ResearchGate discussions. In I. Vázquez and I. Guillén (eds) *ELF and Intercultural Communication: Implications and/or Applications to the Field of English Language Teaching* (pp. 273–302). Bern: Peter Lang.

Luzón, M.J. (2019a) English as a lingua franca in academic blogs: Its co-existence and interaction with other languages. In Z. Tatsioka, B. Seidlhofer, N. Sifakis and G. Ferguson (eds) *Using English as a Lingua Franca in Education in Europe* (pp. 125–149). Berlin: de Gruyter Mouton.

Luzón, M.J. (2019b) Bridging the gap between experts and publics: The role of multimodality in disseminating research in online videos. *Ibérica, Journal of the European Association of Languages for Specific Purposes* 37, 167–192.

Luzón, M.J. (2019c) 'Meet our group!' Addressing multiple audiences on the websites of Spanish Research Groups. *IJES: International Journal of English Studies* 19 (2), 37–59.

Luzón, M.J. and Pérez-Llantada, C. (2019) Connecting traditional and new genres: Trends and emerging themes. In M.J. Luzón and C. Pérez-Llantada (eds) *Science Communication on the Internet: Old Genres Meet New Genres* (pp. 1–18). Amsterdam: John Benjamins.

Luzón, M.J. and Albero, S. (2020) 'Had a lovely week at #conference2018': An analysis of interaction through conference tweets. *RELC Journal* 51 (1), 33–51.

Macdonald, H., Loder, E. and Abbasi (2020) Living systematic reviews at The BMJ. *BMJ* 370: m2925. See https://www.bmj.com/content/370/bmj.m2925. (accessed 20 April 2021).

Mackenzie Owen, J.S. (2007) *The Scientific Article in the Age of Digitization*. Dordrecht: Springer.

Mahrt, M. and Puschmann, C. (2014) Science blogging: An exploratory study of motives, styles, and audience reactions. *Journal of Science Communication* 13 (3), A05. https://doi.org/10.22323/2.13030205. (accessed 20 April 2021).

Manca, S. and Ranieri, M. (2016) 'Yes for sharing, no for teaching!': Social media in academic practices. *Internet and Higher Education* 29, 63–74.

Manco-Vega, A. (2017) Early career researchers and PhD students from the social sciences use of Social Networking Sites (SNS) for science communication: An affordances approach. Master thesis, Uppsala University. See http://uu.diva-portal.org/smash/get/diva2:1151154/FULLTEXT01.pdf (accessed 20 April 2021).

Mandavilli, A. (2011) Trial by Twitter. *Nature* 469 (7330), 286–287.

Marshall, D., Barbour, K. and Moore, C (2017) Academic persona: The construction of online reputation in the modern academy. In D. Lupton, I. Mewburn and P. Thomson (eds) *The Digital Academic: Critical Perspectives on Digital Technologies in Higher Education* (pp. 47–62). Abingdon: Routledge.

Marwick, A. and boyd, d.m. (2011) I tweet honestly, I tweet passionately: Twitter users, context collapse, and the imagined audience. *New Media & Society* 13, 114–133.

Mauranen, A. (2013) Hybridism, edutainment, and doubt: Science blogging finding its feet. *Nordic Journal of English Studies* 13 (1), 7–36.

Mauranen, A. (2018) Conceptualising ELF. In J. Jenkins, W. Baker and M. Dewey (eds) *The Routledge Handbook of English as a Lingua Franca* (pp. 7–24). London: Routledge.

Mauranen, A., Pérez-Llantada, C. and Swales, J.M. (2020) Academic Englishes: A standardised knowledge? In A. Kirkpatrick (ed.) *The Routledge Handbook of World Englishes* (2nd edn; pp. 659–676). London: Routledge.

McGrath, L. (2014) Parallel language use in academic and outreach publication: A case study of policy and practice. *Journal of English for Academic Purposes* 13, 5–16.

McGrath, L. (2016) Open-access writing: An investigation into the online drafting and revision of a research article in pure mathematics. *English for Specific Purposes* 43, 25–36.

Mehlenbacher, A.R. (2017) Crowdfunding science: Exigencies and strategies in an emerging genre of science communication. *Technical Communication Quarterly* 26, 127–144. doi:10.1080/10572252.2017.1287361

Mehlenbacher, A.R. (2019a) Registered reports: An emerging scientific research article genre. *Written Communication* 36 (1), 38–67.

Mehlenbacher, A.R. (2019b) *Science Communication Online: Engaging Experts and Publics on the Internet.* Columbus, OH: The Ohio State University Press.

Mehlenbacher, A.R. and Mehlenbacher, B. (2019) The case of the scientific research article and lessons concerning genre change online. In M.J. Luzón and C. Pérez-Llantada (eds) *Science Communication on the Internet: Old Genres Meet New Genres* (pp. 41–57). Amsterdam: John Benjamins.

Meishar-Tal, H. and Pieterse, E. (2017) Why do academics use academic social networking sites? *The International Review of Research in Open and Distributed Learning* 18 (1). doi:10.19173/irrodl.v18i1.2643 (accessed 20 April 2021).

Menendez, M., de Angeli, A. and Menestrina, Z. (2012) Exploring the virtual space of academia. *Proceedings of the 10th International Conference on the Design of Cooperative Systems.* Marseille, France, 30 May–1 June (pp. 49–63). London: Springer Verlag.

Merchant, G. (2006) Identity, social networks and online communication. *E-Learning* 3, 235–244.

Mewburn, I. and Thomson, P. (2013) Why do academics blog? An analysis of audiences, purposes and challenges. *Studies in Higher Education* 38 (8), 1105–1119.

Mewburn, I. and Thomson, P. (2018) Towards an academic self? Blogging during the doctorate. In D. Lutpon, P. Thomson and I. Mewburn (eds) *The Digital Academic: Critical Perspectives on Digital Technologies in Higher Education* (pp. 20–35). Routledge: London.

Miller, C.R. (1984) Genre as social action. *Quarterly Journal of Speech* 70 (2), 151–167.

Miller, C.R. (2015a) Genre as social action (1984). Revisited 30 Years Later (2014). *Letras & Letras* 31 (3), 56–72.

Miller, C.R. (2015b) Genre change and evolution. In N. Artemeva and A. Freedman (eds) *Genre Studies around the Globe: Beyond the Three Traditions* (pp. 154–185). Edmonton, AB: Inkshed Publications.

Miller, C.R. and Shepherd, D. (2009) Questions for genre theory from the blogosphere. In J. Giltrow and D. Stein (eds) *Genres in the Internet. Issues in the Theory of Genre* (pp. 263–290). Amsterdam: John Benjamins.

Miller, C.R. and Fahnestock, J. (2013) Genres in scientific and technical rhetoric. *Poroi* 9, 1: Article 12. https://doi.org/10.13008/2151-2957.1161 (accessed 20 April 2021).

Miller, C.R., Devitt, A.J. and Gallagher, V.J. (2018) Genre: Permanence and change. *Rhetoric Society Quarterly* 48 (3), 269–277.

Minol, K., Spelsberg, G., Schulte, E. and Morris, N. (2007) Portals, blogs and co.: The role of the Internet as a medium of science communication. *Biotechnology Journal* 2 (9), 1129–1140.

Muñoz-Morcillo, J., Czurda, K. and Robertson-von Trotha, C.Y. (2016) Typologies of the popular science web video. *Journal of Science Communication* 15, 4. See https://jcom.sissa.it/sites/default/files/documents/JCOM_1504_2016_A02.pdf (accessed 20 April 2021).

Muresan, L.M. and Pérez-Llantada, C. (2014) English for research publication and dissemination in bi-/multiliterate environments: The case of Romanian academics. *Journal of English for Academic Purposes* 13, 53–64.

Nández, G. and Borrego, A. (2013) Use of social networks for academic purposes: A case study. *The Electronic Library* 31, 781–791.

Nature Chemistry (2011) Editorial: The art of abstracts. *Nature Chemistry* 3, 571.

Nentwich, M. and Konig, R. (2014) Academia goes Facebook? The potential of social network sites in the scholarly realm. In S. Bartling and S. Friesike (eds) *Opening Science* (pp. 107–124). Cham: Springer International Publishing.

Nielsen, M. (2012) *Reinventing Discovery: The New Era of Networked Science*. Princeton, NJ: Princeton University Press.

Nobarany, S. and Booth K.S. (2015) Use of politeness strategies in signed open peer review. *Journal of the Association for Information Science & Technology* 66 (5), 1048–64.

Norman, M.E. (2000) Public education through community-based film programs: A report on the environmental film festival in the nation's capital. *The Journal of Environmental Education* 31 (2), 28–30.

Nygaard, L.P. and Bellanova, R. (2018) Lost in quantification: Scholars and the politics of bibliometrics. In M.J. Curry and T. Lillis (eds) *Global Academic Publishing: Policies, Perspectives and Pedagogies* (pp. 23–36). Bristol: Multilingual Matters.

O'Halloran, K.L. (2009) Historical changes in the semiotic landscape: From calculation to computation. In J. Carey (ed.) *The Routledge Handbook of Multimodal Analysis* (pp. 98–113). London: Routledge.

O'Halloran, K.L., Tan, S. and Smith, B.A. (2016) Multimodal approaches to English for academic purposes. In K. Hyland and P. Shaw (eds) *The Routledge Handbook of English for Academic Purposes* (pp. 257–270). London: Routledge.

OECD (2015) *Making Open Science a Reality, OECD Science, Technology and Industry Policy Papers*. Paris: OECD Publishing (No. 25).

Oransky, I. and Marcus, A. (2020a) Quick retraction of a faulty coronavirus paper was a good moment for science. *Retraction Watch*, blogpost, 3 February. See https://www.statnews.com/2020/02/03/retraction-faulty-coronavirus-paper-good-moment-for-science/ (accessed 19 April 2020).

Oransky, I. and Marcus, A. (2020b) Nature Communications retracts much-criticized paper on mentorship. *Retraction Watch*, blogpost, 21 December. See https://retractionwatch.com/2020/12/21/nature-communications-retracts-much-criticized-paper-on-mentorship/ (accessed 19 April 2021).

Orlikowski, W. and Yates, J. (1994) Genre repertoire: The structuring of communicative practices in organizations. *Administrative Science Quarterly* 39 (4), 541–574.

Orpin, D. (2019) #Vaccineswork: Recontextualizing the content of epidemiological reports on Twitter. In M.J. Luzón and C. Pérez-Llantada (eds) *Science Communication on the Internet: Old Genres Meet New Genres* (pp. 173–194). Amsterdam: John Benjamins.

Paltridge, B. (2017) *The Discourse of Peer Review: Reviewing Submissions to Academic Journals*. London: Palgrave MacMillan.

Park, H., Rodgers, S. and Stemmle, J. (2013) Analyzing health organizations' use of Twitter for promoting health literacy. *Journal of Health Communication* 18 (4), 410–425.

Pasquali, M. (2007) Video in science. *EMBO Reports* 8 (8), 712–716.

Paulus, T.M. and Roberts, K.R. (2018) Crowdfunding a 'real-life superhero': The construction of worthy bodies in medical campaign narratives. *Discourse, Context & Media* 21, 64–72.

Penn, L. (2018) Alternative ways of obtaining scholarly articles and the impact on traditional publishing models from a UK/European perspective. *Serials Review* 44 (1), 40–50. doi:10.1080/00987913.2018.1433906.

Pérez-Llantada, C. (2012) *Scientific Discourse and the Rhetoric of Globalization. The Impact of Culture and Language*. London: Continuum.

Pérez-Llantada, C. (2013) The article of the future: Strategies for genre stability and change. *English for Specific Purposes* 32 (4), 221–235.

Pérez-Llantada, C. (2016) How is the digital medium shaping research genres? Some cross-disciplinary trends. *ESP Today* 4 (1), 22–42.

Pérez-Llantada, C. (2018) Bringing into focus multilingual realities: Faculty perceptions of academic languages on campus. *Lingua* 212, 30–43.

Pérez-Llantada, C. (2021a) Grammar and discourse style features in digital genres: The case of science-focused crowdfunding projects. *Revista Signos. Estudios de Lingüística* 54 (105), 73–96.

Pérez-Llantada, C. (2021b) *Research Genres across Languages. Multilingual Communication Online*. Cambridge: Cambridge University Press.

Pérez-Llantada, C. (2021c) Genres and languages in science communication: The multiple dimensions of the science-policy interface. *Language & Communication* 78, 65–75. doi:10.1016/j.langcom.2021.02.004

Pérez-Llantada, C. and Swales, J.M. (2017) English for academic purposes. In E. Hinkel (ed.) *Handbook of Research in Second Language Teaching and Learning III* (pp. 42–55). New York: Routledge.

Pham, D. (2016) Public engagement is key for the future of science research. *Science of Learning* 1, 16010. doi:10.1038/npjscilearn.2016.10 (accessed 20 April 2021).

Pigg, S., Hart-Davidson, W., Grabill, J. and Ellenbogen, K. (2016) Why people care about chickens and other lessons about rhetoric, public science, and informal learning environments. In J. Buehl and A. Gross (eds) *Science and the Internet: Communicating Knowledge in a Digital Age* (pp. 267–282). Amityville, NY: Baywood Publishing.

Plastina, A.F. (2017) Professional discourse in video abstracts: Re-articulating the meaning of written research article abstracts. In G. Garzone, P. Catenaccio, K. Grego and R. Doerr (eds) *Specialised and Professional Discourse across Media and Genres* (pp. 57–74). Milano: Ledizioni.

Politzer-Ahles, S., Holliday, J.J., Girolamo, T., Spychalska, M. and Berkson, K.H. (2016) Is linguistic injustice a myth? A response to Hyland. *Journal of Second Language Writing* 34, 3–8.

Pöschl, U. (2012) Multi-stage open peer review: Scientific evaluation integrating the strengths of traditional peer review with the virtues of transparency and self-regulation. *Frontiers in Computational Neuroscience* 6 (33). doi:10.3389/fncom.2012.00033 (accessed 20 April 2021).

Prem, E., Sanz, F.S., Lindorfer, M., Lampert, D. and Irran, J. (2016) *ODS. Open Digital Science. (SMART 2014/0007). Final Study Report*. Vienna: Eutema. See http://ec.europa.eu/newsroom/dae/document.cfm?doc_id=16018. (accessed 20 April 2021).

Puschmann, C. (2009) Lies at Wal-Mart. Style and the subversion of genre in the Life at WalMart blog. In J. Giltrow and D. Stein (eds) *Genres in the Internet* (pp. 49–84). Amsterdam: John Benjamins.

Puschmann, C. (2014) (Micro)blogging science? Notes on potentials and constraints of new forms of scholarly communication. In S. Friesike and S. Bartling (eds) *Opening Science* (pp. 89–106). New York: Springer.

Puschmann, C. (2015a) A digital mob in the ivory tower? Context collapse in scholarly communication online. In M. Bondi, S. Cacchiani and D. Mazzi (eds) *Discourse in and through the Media* (pp. 22–45). Newcastle: Cambridge Scholars.

Puschmann, C. (2015b) The form and function of quoting in digital media. *Discourse, Context & Media* 7, 28–36.

Quinn, C. (2014, November 12) This new scientific journal is kind of like YouTube, but without the cats. *PRI's The World*. See https://www.pri.org/stories/2014-11-14/new-scientific-journal-kind-youtube-without-cats (accessed 24 July 2021).

Raffaghelli, J.E. and Manca, S. (2018) Exploring the geographies of academic social network sites from a socio-technical perspective: An investigation of scientific literature in Spanish. In M. Bajić, N.B. Dohn, M. de Laat, P. Jandrić and T. Ryberg (eds) *Proceedings of the 11th International Conference on Networked Learning 2018* (pp. 409–416). Zagreb: University of Applied Sciences.

Ranger, M. and Bultitude, K. (2016) 'The kind of mildly curious sort of science interested person like me': Science bloggers' practices relating to audience recruitment. *Public Understanding of Science* 25, 361–378.

Regarding mentorship (2020) *Nature Communications* 11, 6447. https://doi.org/10.1038/s41467-020-20618-x (accessed 24 July 2021).

Reid, G. (2017a) Digital writing in the disciplines: Imagining possibilities for scholarship in the context of digital media. PhD thesis, NC State University. http://www.lib.ncsu.edu/resolver/1840.20/34455. (accessed 20 April 2021).

Reid, G. (2017b) Shifting networks of science: Citizen science and scientific genre change. In H. Yu and K. Northcut (eds) *Scientific Communication: Practices, Theories, and Pedagogies* (pp. 19–38). New York: Routledge.

Reid, G. (2019) Compressing, expanding, and attending to scientific meaning: Writing the semiotic hybrid of science for professional and citizen scientists. *Written Communication* 36 (1), 68–98.

Reid, G. and Anson, C.M. (2019) Public- and expert-facing communication: A case study of polycontextuality and context collapse in Internet-mediated citizen science. In M.J. Luzón and C. Pérez-Llantada (eds) *Science Communication on the Internet: Old Genres Meet New Genres* (pp. 219–238). Amsterdam: John Benjamins.

Rentier, B. (2016) Open science: A revolution in sight? *Interlending and Document Supply* 44 (4), 155–160.

Ritson, S. (2016) 'Crackpots' and 'active researchers': The controversy over links between arXiv and the scientific blogosphere. *Social Studies of Science* 46 (4), 607–628.

Ross, J.S., Gross, C.P., Desai, M.M., Hong, Y., Grant, A.O., Daniels, S.R., Hachinski, V.C., Gibbons, R.J. Gardner, T.J. and Krumholz, H.M. (2006) Effect of blinded peer SI review on abstract acceptance. *JAMA* 295 (14), 1675–1680.

Ross-Hellauer, T. (2017) What is open peer review? A systematic review. *F1000Research*. doi:10.12688/f1000research.11369.2 (accessed 20 April 2021).

Rourke, L., Anderson, T., Garrison, D.R. and Archer, W. (1999) Assessing social presence in asynchronous, text-based computer conferencing. *Journal of Distance Education* 14 (3), 51–70.

Rowley-Jolivet, E. and Campagna, S. (2011) From print to Web 2.0: The changing face of discourse for special purposes. *LSP Journal* 2 (2), 44–51.

Rowley-Jolivet, E. and Carter-Thomas, S. (2019) Scholarly soundbites: Audiovisual innovations in digital science and their implications for genre evolution. In M.J. Luzón and C. Pérez-Llantada (eds) *Science Communication on the Internet: Old Genres Meet New Genres* (pp. 81–106). Amsterdam: John Benjamins Publishing Company.

Royal Society (2011) *Knowledge Networks and Nations: Global Scientific Communication in the 21st Century*. London: The Royal Society.

Rozycki, W. and Johnson, N. (2013) Non-canonical grammar in Best Paper award winners in engineering. *English for Specific Purposes* 32 (3), 157–169.

Rudolf von Rohr, M.T., Thurnherr, F. and Locher, M. (2019) Linguistic expert creation in online health practices. In P. Bou-Franch and P. Garcés-Conejos Blitvich (eds) *Analyzing Digital Discourse: New Insights and Future Directions* (pp. 219–250). Cham: Palgrave Macmillan.

Sajeev, E.P.M., Mintz-Woo, K., Damert, M., Brunner, L. and Eise, J. (2019) Blogging climate change: A case study. In F.W. Leal, B. Lackner and H. McGhie (eds) *Addressing the Challenges in Communicating Climate Change Across Various Audiences* (pp. 129–142). Cham: Springer.

Salager-Meyer, F. (2008) Scientific publishing in peripheral (a.k.a. developing) countries: Challenges for the future. *Journal of English for Academic Purposes* 7 (2), 121–132.

Salager-Meyer, F. (2014) Writing and publishing in peripheral scholarly journals: How to enhance the global influence of multilingual scholars? *Journal of English for Academic Purposes* 13, 78–82.

Salager-Meyer, F. (2018) Open access: The next model for research dissemination. In M.J. Curry and T. Lillis (eds) *Global Academic Publishing: Policies, Perspectives and Pedagogies* (pp. 184–199). Bristol: Multilingual Matters.

Santos, Boaventura de Sousa (2002) The processes of globalisation. *Revista Crítica de Cięncias Sociais*, 1–53. See https://www.eurozine.com/the-processes-of-globalisation/.

Sarcina, A. (2019) Open Science: A review of definitions with a regional perspective. *Impakter*, December 6. See https://impakter.com/open-science-a-review-of-definitions-with-a-regional-perspective/

Saunders, M.E., Duffy, M.A., Heard, S.B., Kosmala, M., Leather. S.R., McGlynn, T.P., Ollerton, J. and Parachnowitsch, A.L. (2017) Bringing ecology blogging into the scientific fold: Measuring reach and impact of science community blogs. *R. Soc. Open Sci.* 4, 170957. http://dx.doi.org/10.1098/rsos.170957 (accessed 20 April 2021).

Scanlon, E. (2014) Scholarship in the digital age: Open educational resources, publication and public engagement. *British Journal of Educational Technology* 45 (1), 12–23.

Schallert, D. L., Chiang, Y.V., Park, Y., Jordan, M.E., Lee, H., Cheng, A.J., Chu, H.R., Lee, S., Kim, T. and Song, K. (2009) Being polite while fulfilling different discourse functions in online classroom discussion. *Computers & Education* 53, 713–725.

Schapira, M. and Harding, R. (2019) Open laboratory notebooks: Good for science, good for society, good for scientists. *F1000Research* 8 (87). 10.12688/f1000research.17710.1. (accessed 20 April 2021).

Schekman, R., Watt, F. and Weigel, D. (2013) Scientific publishing: The eLife approach to peer review. *eLife*. 2: e00799. doi:10.7554/eLife.00799 (accessed 20 April 2021).

Scheliga, K., Friesike, S., Puschmann, C. and Fecher, B. (2016) Setting up crowd science projects. *Public Understanding of Science* 27 (5), 515–534.

Science Europe (2017) The Rationales of Open Science: Digitalisation and Democratisation in Research. Science Europe High-Level Workshop, 14 September 2017, Berlin.

Scotto di Carlo, G. (2014) The role of proximity in online popularizations: The case of TED Talks. *Discourse Studies* 16 (5), 591–606.

Seargeant, P., Tagg, C. and Ngampramuan, W. (2012) Language choice and addressivity strategies in Thai-English social network interactions. *Journal of Sociolinguistics* 16 (4), 510–531.

Sebastian, V. and Giménez, M. (2016) Teaching nanoscience and thinking nano at the macroscale: Nanocapsules of wisdom. *Procedia – Social and Behavioral Sciences* 228, 489–495.

Segado-Boj, F., Díaz-Campo, J., Fernández-Gómez, E. and Chaparro-Domínguez, M.A. (2019) Spanish academics and social networking sites: Use, non-use, and the perceived advantages and drawbacks of Facebook, Twitter, LinkedIn, ResearchGate, and Academia.edu. *First Monday* 24 (5). https://doi.org/10.5210/fm.v24i5.7296 (accessed 20 April 2021).

Seidlhofer, B. (2011) *Understanding English as a Lingua Franca*. Oxford: Oxford University Press.

Shah, N. (2015) Factors influencing academics' use of microblogging in Higher Education. PhD thesis, University of Sheffield. http://etheses.whiterose.ac.uk/id/eprint/11505 (accessed 20 April 2021).

Shanahan, M.C. (2011) Science blogs as boundary layers: Creating and understanding new writer and reader interactions through science blogging. *Journalism* 12 (7), 903–919.

Sharevski, F., Jachim, P. and Florek, K. (2020) To tweet or not to tweet: Covertly manipulating a Twitter debate on vaccines using malware-induced misperceptions. https://arxiv.org/abs/2003.12093 (accessed 20 April 2021).

Shema, H., Bar-Ilan and J. Thelwall, M. (2012) Research blogs and the discussion of scholarly information. *PLoS One* 7, e35869. doi:10.1371/journal.pone.0035869 (accessed 20 April 2021).

Shema, H., Bar-Ilan, J. and Thelwall, M. (2015) How is research blogged? A content analysis approach. *Journal of the Association for Information Science and Technology* 66 (6), 1136–1149.

Shepherd, M. and Watters, C. (1998) The evolution of cybergenres. *Proceedings of the 31st Annual Hawaii International Conference on Systems Sciences (HICSS-31), Vol. II* (pp. 97–193). Los Alamitos: IEEE Computer Society Press.

Shiffman, D.S. (2012) Twitter as a tool for conservation education and outreach: What scientific conferences can do to promote live-tweeting. *Journal of Environmental Studies and Sciences* 2, 257–262.

Sidler, M. (2016) The chemistry liveblogging event: The web refigures peer review. In A. Gross and J. Buehl (eds) *Science and the Internet: Communicating Knowledge in a Digital Age* (pp. 99–116). Amityville, NY: Baywood Publishing.

Sitek, D. and Bertelmann, R. (2014) Open access: A state of the art. In S. Bartling and S. Friesike (eds) *Opening Science* (pp. 139–143). Cham: Springer.

Skutnabb-Kangas, T. and Phillipson, R. (2011) Language ecology. In J. Verschueren, J.O. Östman, J. Blommaert and C. Bulcaen (eds) *Handbook of Pragmatics* (pp. 1–18). Amsterdam: John Benjamins.

Smith, A. (2015) 'Wow, I didn't know that before; thank you': How scientists use Twitter for public engagement. *Journal of Promotional Communications* 3 (3), 320–339.

Snyder, I. (eds) (1998) *Page to Screen – Taking Literacy into the Electronic Era*. London: Routledge.

Soete, L. Schneegans, S., Eröcal, D. Angathevar, B. and Rasiah, R. (2015) A world in search of an effective growth strategy. In *UNESCO Science Report: Towards 2030* (pp. 20–55). See https://en.unesco.org/sites/default/files/usr15_a_world_in_search_of_an_effective_growth_strategy.pdf (accessed 20 April 2021).

Sokół, M. (2018) 'Have you wondered why sportspeople die?' The medical weblog as a popularisation tool. *Discourse, Context & Media* 25, 13–24.

Spanish Strategy for Science and Technology and Innovation (2012) *Spanish Ministry of Economy and Competitiveness*. See https://www.ciencia.gob.es/stfls/MICINN/Investigacion/FICHEROS/Spanish_Strategy_Science_Technology.pdf (accessed 20 July 2017).

Spicer, S. (2014) Exploring video abstracts in science journals: An overview and case study. *Journal of Librarianship and Scholarly Communication* 2 (2), eP1110. http://dx.doi.org/10.7710/2162-3309.1110 (accessed 20 April 2021).

Spinuzzi, C. (2003) *Tracing Genres through Organizations: A Sociocultural Approach to Information Design*. Cambridge, MA: MIT Press.

Spinuzzi, C. (2004) Four ways to investigate assemblages of texts: Genre sets, systems, repertoires, and ecologies. *ACM SIGDOC 2004 Conference Proceedings*. New York: ACM. https://dl.acm.org/doi/10.1145/1026533.1026560 (accessed 20 April 2021).

Spinuzzi, C. and Zachry, M. (2000) Genre ecologies: An open-system approach to understanding and constructing documentation. *ACM Journal of Computer Documentation* 24 (3), 169–181.

Stewart, B. (2015a) Scholarship in abundance: Influence, engagement, and attention in scholarly networks. PhD thesis, University of Prince Edward Island. See http://bonstewart.com/Scholarship_in_Abundance.pdf (accessed 20 April 2021).

Stewart, B. (2015b) Open to influence: What counts as academic influence in scholarly networked Twitter participation. *Learning, Media and Technology* 40 (3), 287–309.

Sugimoto C.R., Work, S., Larivière, V. and Haustein, S. (2017) Scholarly use of social media and altmetrics: A review of the literature. *Journal of the Association for Information Science and Technology* 68, 2037–2062.

Swales, J.M. (1990) *Genre Analysis. English in Academic and Research Settings*. Cambridge: Cambridge University Press.

Swales, J.M. (2004) *Research Genres: Explorations and Applications*. Cambridge: Cambridge University Press.

Swales, J.M. (2019a) Envoi. In J. Corcoran, K. Englander and L.M. Muresan (eds) *Pedagogies and Policies for Publishing Research in English: Local Initiatives Supporting International Scholars* (pp. 284–290). New York: Routledge.

Swales, J.M. (2019b) The futures of EAP genre studies: A personal viewpoint. *Journal of English for Academic Purposes* 38, 75–82.

Szabla, M. and Blommaert, J. (2020) Does context really collapse in social media interaction? *Applied Linguistics Review* 11 (2), 251–279.

Tagg, B. (2015) *Exploring Digital Communication: Language in Action*. London: Routledge.

Tagg, C. and Seargeant, P. (2012) Writing systems at play in Thai-English online interactions. *Writing Systems Research* 4 (2), 195–213.

Tattersall, A. (2015) For what it's worth: The open peer review landscape. *Online Information Review* 39. See https://doi.org/10.1108/OIR-06-2015-0182 (accessed 20 April 2021).

Taylor & Francis Group (2015) Peer review in 2015: A global view. See https://authorservices.taylorandfrancis.com/publishing-your-research/peer-review/peer-review-global-view/ (accessed 20 April 2021).

Teixeira da Silva, J.A. (2016) Vigilantism in science: The need and the risks. *Academic Journal of Interdisciplinary Studies* 5 (3), 9. See http://www.richtmann.org/journal/index.php/ajis/article/view/9543 (accessed 24 July 2021).

Tennant, J.P., Dugan, J.M., Graziotin, D. *et al.* (2017) A multi-disciplinary perspective on emergent and future innovations in peer review. *F1000Research* 6: 1151. doi:10.12688/f1000research.12037.3 (accessed 20 April 2021).

Thelwall, M. and Kousha, K. (2015) ResearchGate: Disseminating, communicating, and measuring scholarship? *Journal of the Association for Information Science and Technology* 66 (5), 876–889.

Thorsen, E. (2013) Blogging on the ice: Connecting audiences with climate-change sciences. *International Journal of Media and Cultural Politics* 9 (1), 87–101.

Torny, D. (2018) Pubpeer: Vigilante science, journal club or alarm raiser? The controversies over anonymity in post-publication peer review. PEERE International Conference on Peer Review, PEERE COST Network, March 2018, Rome, Italy. https://hal.inria.fr/halshs-01700198/ (accessed 24 July 2021).

Townsend, F. (2013) Post-publication peer review: PubPeer. *Editors' Bulletin* 9, 45–46. http://doi.org/10.1080/17521742.2013.865333 (accessed 20 April 2021).

Trench, B. (2008) Internet: Turning science communication inside out? In M. Bucchi and B. Trench (eds) *Handbook of Public Communication of Science and Technology* (pp. 185–198). London: Routledge.

Trench, B. (2012) Scientists' blogs: Glimpses behind the scenes. In S. Rodder, M. Franzen and P. Weingart (eds) *The Sciences' Media Connection: Public Communication and its Repercussions* (pp. 273–289). Dordrecht: Springer.

Tribble, C. (2017) ELFA vs. Genre: A new paradigm war in EAP writing instruction? *Journal of English for Academic Purposes* 25, 30–44.

Trotter, H., Kell, C., Willmers, M., Grey, E. and King, T. (2014) *Seeking Impact and Visibility: Scholarly Communication in Southern Africa*. Cape Town: African Minds.

Tusting, K. (2018) The genre regime of research evaluation: Contradictory systems of value around academics' writing. *Language and Education* 32 (6), 477–493.

Vale, R.D. and Hyman, A.A. (2016) Priority of discovery in the life sciences. *eLife* 5, e16931. doi: 10.7554/eLife.16931.001 (accessed 20 April 2021).

Van Leeuwen, T. (2005) Multimodality, genre and design. In R. Jones and S. Norris (eds) *Discourse in Action – Introducing Mediated Discourse Analysis* (pp. 73–95). London: Routledge.

Veletsianos, G. (2012) Higher education scholars' participation and practices on Twitter. *Journal of Computer Assisted Learning* 28, 336–349.

Veletsianos, G. (2013) Open practices and identity: Evidence from researchers and educators' social media participation. *British Journal of Educational Technology* 44 (4), 639–651.

Veletsianos, G. and Kimmons, R. (2012) Networked participatory scholarship: Emergent techno-cultural pressures toward open and digital scholarship in online networks. *Computers and Education* 58(2), 766–774. http://doi.org/10.1016/j.compedu.2011.10.001

Veletsianos, G. and Kimmons, R. (2013) Scholars and faculty members lived experiences in online social networks. *The Internet and Higher Education* 16 (1), 43–50.

Vicente-Sáez, R. and Martínez-Fuentes, C. (2018) Open science now: A systematic literature review for an integrated definition. *Journal of Business Research* 88, 428–436.

Vitak, J. (2012) The impact of context collapse and privacy on social network site disclosures. *Journal of Broadcasting and Electronic Media* 56, 451–470.

Walbot, V. (2009) Are we training pit bulls to review our manuscripts? *Journal of Biology* 8, 24. https://doi.org/10.1186/jbiol125 (accessed 24 July 2021).

Wall, A. and Spinuzzi, C. (2018) The art of selling-without-selling: Understanding the genre ecologies of content marketing. *Technical Communication Quarterly* 27 (2), 137–160.

Walsh, J.P. and Roselle, A. (1999) Computer networks and the virtual college. *Science Technology Industry Review* (OECD) (24), 49–78.

Walter, S., Lörcher, I. and Brüggemann, M. (2019) Scientific networks on Twitter: Analyzing scientists' interactions in the climate change debate. *Public Understanding of Science* 28 (6), 696–712.

Weller, M. (2011) *The Digital Scholar: How Technology is Transforming Scholarly Practice*. Basingstoke: Bloomsbury Academic.

Weller, K., Dröge, E. and Puschmann, C. (2011) Citation analysis in Twitter: Approaches for defining and measuring information flows within tweets during scientific conferences. In M. Rowe, M. Stankovic, A.S. Dadzie and M. Hardey (eds) *#MSM* (pp. 1–12). See CEUR-WS.org.

Wendel, J.N. (2005) Notes on the ecology of language. *Bunkyo Gakuin University Academic Journal* 5, 51–76.

Wernli, D. and Darbellay, F. (2016) *Interdisciplinarity and the 21st Century Research-Intensive University*. Report for the League of European Research Universiti (LERU). See https://www.leru.org/publications/interdisciplinarity-and-the-21st-century-research-intensive-university (accessed 20 April 2021).

Wickman, Ch. (2016) Learning to 'share your science': The Open Notebook as textual object and dynamic rhetorical space. In A. Gross and J. Buehl (eds) *In Science and the Internet: Communicating Knowledge in a Digital Age* (pp. 11–22). London: Routledge.

Wilkins, J.S. (2008) The roles, reasons and restrictions of science blogs. *Trends in Ecology & Evolution* 23 (8), 411–413.

Wittgenstein, L. (1953) *Philosophical Investigations*. Oxford: Blackwell Publishing.

Wolfe-Simon, F., Switzer Blum, J., Kulp, T.R., Gordon, G.W., Hoeft, S.E., Pett-Ridge, J., Stolz, J.F., Webb, S.M., Weber, P.K., Davies, P.C.W., Anbar, A.D. and Oremland,

R.S. (2011) A bacterium that can grow by using arsenic instead of phosphorus. *Science* 332 (6034), 1163–1166.

Wynants, L., Van Calster, B., Collins, G.S., Riley, R.D., Heinze, G., Schuit, E. *et al.* (2020) Prediction models for diagnosis and prognosis of covid-19: systematic review and critical appraisal. *BMJ*, 369:m1328. doi:10.1136/bmj.m1328. (accessed 24 July 2021).

Yang, W. (2017) Audioslide presentations as an appendant genre. Key words, personal pronouns, stance and engagement. *ESP Today* 5 (1), 24–45.

Yen, C.P. and Hung, T.W. (2018) New data on the linguistic diversity of authorship in Philosophy journals. *Erkenntnis* 84, 953–974.

Yeo, S.K., Liang, X., Brossard, D., Rose, K.M., Korzekwa, K., Scheufele, D. and Xenos, M. (2017) The case of #arseniclife: Blogs and Twitter in informal peer review. *Public Understanding of Science* 26 (8), 937–952.

Yu, H., Xu, S. and Xiao, T. (2018) Is there lingua franca in informal scientific communication? Evidence from language distribution of scientific tweets. *Journal of Informetrics* 12 (3), 605–617.

Zappavigna, M. (2012) *Discourse of Twitter and Social Media: How we Use Language to Create Affiliation on the Web*. London: Bloomsbury.

Zou, H. and Hyland, K. (2019) Reworking research: Interactions in academic articles and blogs. *Discourse Studies* 21 (6), 713–733.

Index

For Product Safety Concerns and Information please contact our EU Authorised Representative:

Easy Access System Europe

Mustamäe tee 50

10621 Tallinn

Estonia

gpsr.requests@easproject.com

www.ingramcontent.com/pod-product-compliance
Ingram Content Group UK Ltd.
Pitfield, Milton Keynes, MK11 3LW, UK
UKHW021836150926
8656IPUK00003B/13
* 9 7 8 1 7 8 8 9 2 4 7 1 9 *